MONTHLY STAR CHARTS

24 ALL-SKY CHARTS FOR STAR WATCHERS WORLDWIDE

George Lovi
Graham Blow

Edited by Alan M. MacRobert and Stephen Peters
Foreword by Leif J. Robinson

Sky Publishing Corporation
Cambridge, Massachusetts

Cover and text design by Mary Santa Maria

Photograph credits:
Lee C. Coombs, 34, 44; Dennis di Cicco, 40, 44, 48, 52; Akira Fujii, 46, 62; Martin C. Germano, 16, 22, 24, 42, 44, 54, 56, 60, 62; Preston Scott Justis, 20, 22, 24, 26, 28, 30, 32, 50, 52; Evered Kreimer, 14; Juhani Salmi, 28; Chuck Vaughn, 12, 18, 20, 22, 54; George R. Viscome, 16, 34, 58; Mark Ward, 52; Kim Zussman; 32, 48, 54, 58.

Published by:

Sky Publishing Corp.
49 Bay State Rd.
Cambridge, MA 02138-1200
U.S.A.

Library of Congress Cataloging-in-Publication Data

Lovi, George, 1939–
Montly star charts: 24 all-sky charts for star watchers worldwide / George Lovi, Graham Blow; edited by Alan M. MacRobert and Stephen Peters; foreword by Leif J. Robinson.
p. cm.
ISBN 0-933346-69-7: $24.95
1. Stars — Observer's manuals. I. Blow, Graham, 1954– . II. MacRobert, Alan M., 1951– , and Peters, Stephen, 1959– . III. Title.
QB63.L683 1995
523.8'022'3 — dc20 94–40236

Printed in the United States of America

95 96 97 98 99 8 7 6 5 4 3 2 1

Table of Contents

Foreword 4
Introduction 5

The Northern Hemisphere

January *A Bull Chock Full* 12
February *A Tale of Two Stars* 14
March *The Heavenly Picture Book* 16
April *Anatomy of a Great Bear* 18
May *Cosmic Clusters Large and Small* 20
June *The Man with the Snake* 22
July *Legends of Herculean Proportions* 24
August *Milky Way Meanderings* 26
September *Figuring How Far to a Star* 28
October *Following Those Variable Stars* 30
November *Peregrinations with Pegasus* 32
December *A Look at Seeing* 34

The Southern Hemisphere

January *Magellanic "Toolhouses of Great Merit"* 40
February *Great Star of the South* 42
March *The Treasures of Argo* 44
April *The Crux Connection* 46
May *Crosses True and False* 48
June *A Strange Case of Celestial Acne* 50
July *Globulars Galore* 52
August *Something for Everyone in Sagittarius* 54
September *The Scorpion's Sting* 56
October *A Traverse of the Southern Ocean* 58
November *The Southern Sky's Attic* 60
December *Boating Down the Eridanus* 62

Appendix A *The Greek Alphabet* 64
Appendix B *The 88 Constellation Names* 65
Index 66

FOREWORD

THE most boring job I've ever had at *Sky & Telescope* was checking the 24 star charts that George Lovi drafted for the magazine in 1967, and which are now gathered together in this book. They were essentially impeccable. With only a handful of exceptions, every star was plotted perfectly, as if with the aid of some sophisticated computer program. But there was no computer — Lovi painstakingly drew each chart largely from *memory,* with pen and India ink.

Traditional mappers of the heavens such as Lovi and Wil Tirion designed their maps with great empathy toward the sky, a sense that only intimate familiarity with the stars and their lore can bring. That grand perspective, that sensitivity to form and history, may be flagging. These days any graphic artist armed with a Macintosh computer and a database can crank out whatever celestial chart you please, usually jazzed up with lots of colors and fancy graphic elements. It may look impressive, but is it accurate? Intelligently labeled? Is it clear and easy to read under a nighttime sky? Too often, the answer to these questions is no.

Indeed, by today's design standards, Lovi's star charts seem plain. Yet this very simplicity translates into utility in the field. And that's what really counts, whether you're just starting out or you've been scanning the skies for years.

For three decades, as I've traveled the world, I've taken along tear sheets of Lovi's star charts. Their multiple horizons — an innovation by *Sky & Telescope*'s founder and first editor, Charles A. Federer, Jr. — give them an unmatched utility throughout temperate latitudes, both north and south. And they show all the stars that most of us can hope to see with our unaided eyes as we toil under metropolitan skies. Yet there is enough detail to key us to unfamiliar constellation patterns and the wonders awaiting binoculars and small telescopes. Also included on these charts are important markers of celestial orientation such as the ecliptic, celestial equator, meridian, and right ascension and declination.

Don't let the title of this book fool you — there's more here than just the charts. Accompanying them are essays by Lovi and astronomer Graham Blow that put each month's skyscape into perspective. We pored over hundreds of columns from Lovi's Rambling Through the Skies series, which appeared every month for 22 years in *Sky & Telescope,* and we finally selected 16 of them for this book. His knack for storytelling and his profound knowledge of constellation legends and lore really bring the heavens to life. While Lovi concentrated mostly on the Northern Hemisphere sky, Blow's eight essays round out the view from south of the equator. I know you'll enjoy their forays into the history, science, and feature attractions of the skies for each month.

To encourage you to take a closer look at some of those attractions, we've included a table of interesting objects with each map. All of the fine galaxies, nebulae, star clusters, double stars, and variable stars listed in each table are plotted on the accompanying map. Give them a look through a pair of binoculars or, better yet, a telescope — and you'll be hooked!

Since 1968, when George Lovi's star charts first appeared in *Sky & Telescope,* amateur astronomers around the world have used them to learn the names and shapes of the constellations, plan their evening observing sessions, and locate the most dazzling deep-sky curiosities. So it pleases me enormously to see them collected in one place — with supporting text, tables, and photographs — for present and future generations to enjoy.

— Leif J. Robinson
Editor-in-Chief, *Sky & Telescope*

Introduction

How to Use the Star Charts

Many people, when I tell them I work for an astronomy magazine, say they wish they knew the stars and constellations. "Why not learn them?" I ask. "Oh, I'm no good at math," they sometimes answer. Or, "I can't afford a telescope." Or as one woman in her 40s told me, "I flunked geometry in high school."

You don't need math, and certainly not a telescope, to look up on a warm spring evening and say "There's Arcturus" — any more than you need a biology degree to recognize a robin. All you need is a good sky map and knowledge of how to use it.

The maps in this book have introduced untold thousands of people to the constellations, having appeared in every issue of *Sky & Telescope* magazine from 1968 to 1992. More than that, the maps continued to serve many readers for years after they became deeply involved in amateur astronomy. The maps are straightforward enough for the beginner, yet they are richer in useful information than any other seasonal all-sky charts available. And they embody an attention to detail that grows more and more appreciated with continued use.

Step 1: Your Date and Time

The sky is always changing. The stars, like the Sun and the Moon, rise in the east, cross the heavens, and set in the west. And they make these journeys at different times of the night (or day) during different months of the year.

Find the map for the evening hours of the current month. Times are labeled on each map's upper right corner. The times are all standard, not daylight saving, so when "summer time" is in effect, add an hour to the times printed.

You can skywatch later at night than the times listed on the map by simply switching to the next month's map. *One month ahead* equals *two hours ahead* in the clockwork of the heavens. In other words, if it's early January and you want to observe at 11 p.m., use the map that says "Early February, 9 o'clock." The table below tells which chart to use for any date and time, in either the Northern or Southern Hemisphere.

Step 2: Mark Your Latitude

The maps are designed for use in almost any inhabited part of the globe. The first 12 maps are drawn for the world's Northern Hemisphere; the last 12 are for skywatchers south of the equator.

Check the world map on page 7 to find your latitude, if you don't already know it. You don't need to know it accurately; to the nearest 10° will do.

Now look at your chosen star map. Let's assume it's for the Northern Hemisphere. Inside the top and bottom edges are wide arcs, more or less parallel to the edges, labeled "20° Horizon," "30° Horizon," "40° Horizon," and "50° Horizon." Notice that each of these arcs continues around to form a complete circle enclosing most of the map. The circle marked for your latitude represents the horizon that you would see encircling you if you were standing outdoors on an open plain.

All stars outside this circle are below your horizon. Ignore them. You might highlight your latitude circle with a colored marking pen for clarity. (You can interpolate between the lines to draw in your exact horizon to an accuracy of 1°, but for most purposes this won't be needed.)

Which Chart to Use?

Use this table to select the correct star chart for any time of night throughout the year, from either the Northern Hemisphere or the Southern Hemisphere. Choose the current month and time from the boldfaced headings; each entry within the table corresponds to the star chart for that month.

For example, in late March at 10 p.m., use the chart labeled "April." In late April at 4 a.m., use the chart labeled "August." Where no chart is designated, use the one listed for the hour either just before or after the current time.

		6 p.m.	**7 p.m.**	**8 p.m.**	**9 p.m.**	**10 p.m.**	**11 p.m.**	**Midnight**	**1 a.m.**	**2 a.m.**	**3 a.m.**	**4 a.m.**	**5 a.m.**	**6 a.m.**
JANUARY	**Early**		December		January		February		March		April		May	
	Late	December		January		February		March		April		May		June
FEBRUARY	**Early**		January		February		March		April		May		June	
	Late	January		February		March		April		May		June		July
MARCH	**Early**		February		March		April		May		June		July	
	Late	February		March		April		May		June		July		August
APRIL	**Early**		March		April		May		June		July		August	
	Late	March		April		May		June		July		August		September
MAY	**Early**		April		May		June		July		August		September	
	Late	April		May		June		July		August		September		October
JUNE	**Early**		May		June		July		August		September		October	
	Late	May		June		July		August		September		October		November
JULY	**Early**		June		July		August		September		October		November	
	Late	June		July		August		September		October		November		December
AUGUST	**Early**		July		August		September		October		November		December	
	Late	July		August		September		October		November		December		January
SEPTEMBER	**Early**		August		September		October		November		December		January	
	Late	August		September		October		November		December		January		February
OCTOBER	**Early**		September		October		November		December		January		February	
	Late	September		October		November		December		January		February		March
NOVEMBER	**Early**		October		November		December		January		February		March	
	Late	October		November		December		January		February		March		April
DECEMBER	**Early**		November		December		January		February		March		April	
	Late	November		December		January		February		March		April		May

Step 3: Getting Oriented

At the top, bottom, and sides of your map, find the words North, West, South, and East. These mark compass points on your horizon. They indicate which small area of the map to look at when you're facing a given direction.

The center of the map shows the sky straight overhead (the *zenith*). Therefore, a star that's plotted on the map halfway from your horizon to the center can be spotted about *halfway up the sky*. In other words, it's roughly 45° above the horizon — about midway between horizontal and straight up.

This shows, by the way, how shrunken and compressed the map really is. From the horizon to overhead is just a few inches on the chart, but in real life it's a huge sweep of your eye and neck.

Step 4: Your First Star

Let's try it out. Suppose it's 9 p.m. in early January; choose the map for that time and date. Fold the book back on itself so only the map is in front of you. Perhaps your yard has a good view toward the southeast. Go outdoors, face southeast, hold the book in front of you, and turn the map around so that its southeast horizon (halfway between "South" and "East") is *down*. You'll be holding the book at an odd angle.

The stars over the southeast horizon on the map now match the stars in front of you in the sky. Roughly a third of the way from the map's southeast horizon to the center, you'll find a big star dot labeled "Sirius." The dot is larger than any other, meaning Sirius is the brightest star. Look a third of the way up the southeastern sky. There it is!

Now look a little above Sirius on the map (more toward the center). There's a pattern of bright stars labeled "Orion." Look in the sky above Sirius. There's the exact same pattern, twinkling away.

Turn around to face west, and rotate the book so the map's "West" edge is down. Over the western horizon on the map is the Great Square of Pegasus, balancing on one corner. The bottom corner star is named Markab. The Square is a little less than half the distance from the horizon to the center, so look into the darkness a little less than halfway from horizontal to straight up. There's the Great Square. Its star dots are smaller than Sirius and some of those in Orion, so expect the Square to be dimmer.

That's it. You're now equipped to identify all the stars and constellations visible any time of the year from anywhere in the world.

Helpful Hints

Look for the brightest stars first, the ones with the biggest dots on the map. Faint stars are hard or impossible to see if you live in a thickly populated area where the sky is full of *light pollution*. Light pollution is the gray skyglow from all the poorly designed and improperly aimed outdoor light fixtures for dozens of miles around. And even if you have a good, black rural sky, there's a much greater difference between bright and faint stars than the dot sizes suggest.

Remember that constellations are big. They look a lot smaller here on paper than in the sky. To see just how small, look for the Big Dipper on the map. It's in the northern part of the sky on all the Northern Hemisphere charts. The bowl of the Big Dipper covers about as much sky as your fist held at arm's length. Try it. Now you know how big a dipper to look for. For Southern Hemisphere observers, the Southern Cross is about ⅔ of a fist long.

Find the dashed line labeled "Meridian" running through each map's center, from the north to the south horizons. (Outdoors, the meridian is an imaginary line that runs from the north point on the horizon up across the zenith and down to the horizon's south point.) Along this line on the map are little labels marking every 10° of declination: +60°, +70°, and so forth. These labels are a fist-width apart. Knowing this, you can use your fist to measure your way from one star group to the next all the way across the map.

Notice that the declination marks are spaced a little more widely near the horizon than overhead, a consequence of the map's stereographic projection. This accords well with how we actually perceive the sky. The well-known "Moon illusion" effect makes the constellations (as well as the Moon and Sun) look bigger when they're near the horizon than when they're overhead. Of course, it's purely an illusion; the constellations never really change size.

Look out for planets. Planets are not plotted on the charts, because they move from week to week with respect to the background stars. But the planets always stay near the dashed line labeled "Ecliptic." (So do the Sun and Moon.)

If you find a bright "star" close to the ecliptic that's missing from the map, chances are you've happened upon a planet. Another way to tell it's a planet is that it usually does not twinkle, the way a star does. You can find out which planet it is by looking in an astronomy magazine such as *Sky & Telescope* for the current month. Mercury, Venus, Mars, Jupiter, and Saturn are all visible to the naked eye at various, changing times of the year.

Let your eyes dark-adapt. You don't see many stars when you first step outdoors. But more and more pop into view as your eyes adjust to the darkness. It takes at least 20 minutes for your night vision to develop. In truly dark conditions it continues to improve for another hour or more. But most people live under too much light pollution for their dark adaptation ever to become this complete.

Make the most of moonless nights. Bright moonlight washes out faint stars, too. Full Moon is the worst time of the month for skywatchers; you see only the brightest stars. It's best to stargaze on moonless nights, or when the Moon is only a crescent. When a bright Moon is up, expect to see only the brightest stars.

Use a red flashlight. Of course, it's hard to read a star chart in the dark. You need a flashlight, preferably a small one (a penlight). But white light ruins your night vision. Red light is what you want.

A time-honored method of modifying a flashlight to emit red light is to secure a piece of red paper or cellophane over the front lens with a rubber band. Or you can cut out a disk of the material and install it behind the lens. Another trick is to paint the lens with several coats of red fingernail polish. The light getting through should be barely bright enough to read by, no brighter.

An even better solution is to use a red light-emitting diode (LED) instead of a regular flashlight bulb. An LED produces a purer, deeper red color, which has even less effect on dark-adapted eyes. LEDs also draw very little battery power. Several LED astronomer's flashlights are commercially available, or you can make your own with parts from an electronics shop.

Take your time. Don't try to learn too much at once. Add one constellation to your repertoire each clear evening, and in a month you'll know the sky better than at least 95 percent of humanity — and with no math degree.

More About the Charts

As you get deeper into astronomy, here are some more facts about the maps that you may want to know.

Equator

Notice the dashed line labeled "Equator." That's the celestial equator. It divides the celestial sphere into two hemispheres: northern and southern. The celestial equator is simply an imaginary projection of the Earth's equator out into space.

Coordinate Grid

To specify locations in the sky, astronomers use a system of other lines also superimposed on the celestial sphere, similar to the latitude and longitude lines used to designate locations here on Earth. But instead of latitude and longitude, they are called *declination* and *right ascension*.

Declination is simply the distance of an object from the celestial equator. It is measured in degrees, arc minutes, and arc seconds (° , ′, and ″) north (+) or south (–) of the celestial equator (which, like Earth's equator, is defined as 0°). There are 60 arc minutes in one degree, and 60 arc seconds in one arc minute.

Running longitudinally between the north and south poles on the celestial sphere — perpendicular to the lines of declination — are *hour circles,* the lines of right ascension. There are 24 hour circles (or really, half-circles), which are counted eastward from the 0-hour mark at a point called the *vernal equinox,* in the constellation Pisces. Right ascension is expressed in hours, minutes, and seconds of time (h, m, and s).

So, the exact position of any star or deep-sky object is described by its celestial coordinates: right ascension (R.A.) and declination (Dec.). Sirius, for example, can be found blazing away at R.A. $6^h 45^m 09^s$, Dec. +16° 42′ 58″.

On the charts in this book, tick marks along the meridian and just inside the outer horizon circle designate every 10° of declination. Also inside the outer horizon are tick marks indicating right ascension. Hour circles are shown fully at three and six hours from the meridian by dashed lines.

A phenomenon called *precession* causes the orientation of the Earth's axis to change over time with respect to the celestial sphere. This means that the

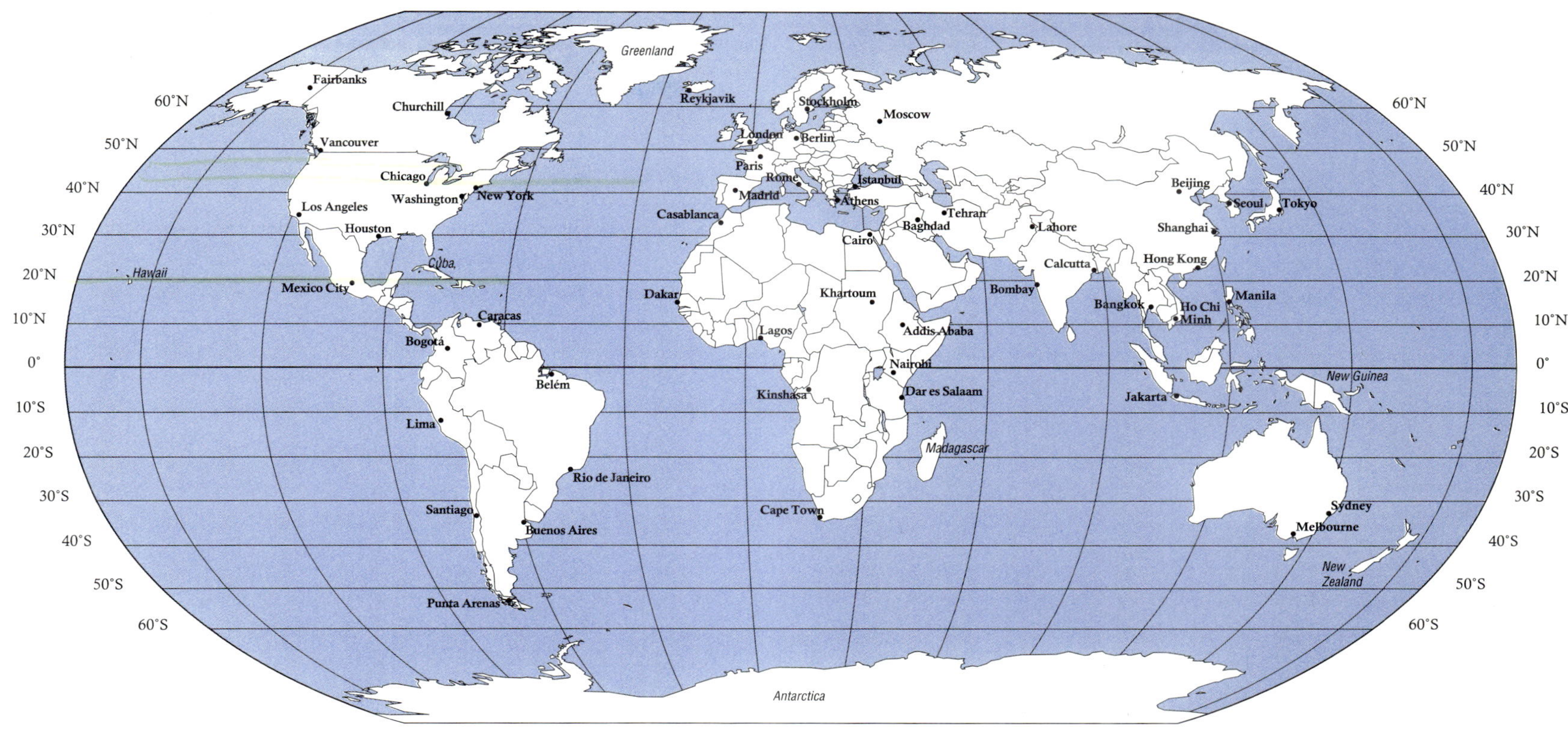

Use this map to determine your latitude to the nearest 10°. Then find the corresponding latitude circle on the star charts — labeled "30° Horizon," "40° Horizon," and so forth — for your hemisphere. Stars outside of that circle are below the horizon and therefore not visible.

coordinate grid shifts gradually with respect to the stars. Thus, every so often their coordinates have to be restated.

The charts in this book, drafted by George Lovi in 1967, show the exact positions of celestial objects for the beginning of 1975. Although precession has been at work, its effect is barely noticeable at the small scale of these charts. For their intended use in naked-eye stargazing, the charts will be sufficiently accurate for many decades to come.

Deep-Sky Objects

Galaxies, nebulae, and open and globular star clusters are indicated on the maps with symbols (defined in the bottom right-hand corner of each map) and a number. Numbers preceded by an M are Messier numbers, identifying objects cataloged in the 18th century by French comet hunter Charles Messier (a few were added by others). Today these 110 assorted deep-sky treasures are known as the *Messier objects.* M45, for example, is the Pleiades, an open star cluster in the constellation Taurus.

Most of the numbers without an M are NGC numbers, assigned to objects in the *New General Catalogue of Nebulae and Clusters of Stars,* published originally in 1888 by Danish astronomer Johann Dreyer. Most Messier objects also have NGC numbers, but on our maps only their Messier numbers are shown. All deep-sky objects visible to the unaided eye or with modest optical aid are included on our charts.

Double and Variable Stars

Prominent *double stars* — including pairs distinguishable with the naked eye — are symbolized on our charts with a line bisecting a dot. Many of what we see as single stars are really two or more stars revolving closely around each other. These double or multiple star systems — as well as so-called optical doubles which by chance appear close together along our line of sight but really are not binary partners — can be interesting to observe, especially when the individual stars have markedly different hues. It can be both fun and challenging to try to resolve the two (or more) stellar components in a telescope.

Most stars shine with a steady intensity, but some vary in brightness. These are called *variable stars.* The more prominent ones are shown on the charts, indicated either by a dot enclosed within a circle, for variables that never dim below naked-eye visibility, or by an open circle, for those that dim to below 6th magnitude or never become brighter than 5th magnitude. George Lovi gives a nice minicourse on variable stars on page 28, and he tells how to monitor their changes in brightness.

The Milky Way

If you're far from city lights on a moonless night, you may see a hazy glowing band arching across the entire dome of the night sky. It's called the Milky Way. This very dense panorama of stars marks the plane of our galaxy, the Milky Way Galaxy. On the star charts, the approximate boundaries of the Milky Way band are indicated by dotted lines. Between them is a dashed line labeled "Galactic Equator," which is the approximate midline of the galactic plane.

Star Magnitudes

The maps show all stars with a visual magnitude of 4.25 or brighter, plus a number of fainter ones to fill out constellations as needed. Magnitude is a measure of how bright a star looks to the eye. The brighter a star, the *lower* its magnitude. The system even extends into negative numbers. For example, Sirius, the brightest star in the sky, has a visual magnitude of –1.4. Slightly less brilliant is Rigel, in the constellation Orion, which is a zero-magnitude star. Polaris, the North Star, is dimmer still, with a magnitude of 2. The faintest you can see with the naked eye in a sky free of light pollution is about 6. In the bottom left corner of each star chart in this book is a magnitude key. As it shows, the dot sizes are scaled in whole-magnitude increments, from "Minus One" to "Fifth."

Star Names and Designations

Every star of 3rd magnitude or brighter is labeled with a Greek letter, a number, or other designation if it has one. Fainter stars that have a Greek letter are so labeled as well. Greek letters were assigned to the brightest stars in each constellation by German astronomer Johann Bayer and others starting in the very early 17th century, and they're still used today. A few stars are labeled with lowercase Roman letters or with numbers from John Flamsteed's star catalog of 1725. The proper names of many stars are also provided.

A star's Greek letter or Flamsteed number is used with the Latin genitive of the constellation name (listed in Appendix A). For example, the Alpha star of Centaurus is Alpha Centauri.

Tables of Interesting Objects

Opposite each monthly star chart for both the Northern and Southern Hemispheres is a table of interesting objects. It contains essential information about particular stars and deep-sky objects ideally positioned for viewing that month in the early evening. While a few of these objects can be seen with the unaided eye, most will be best appreciated when observed through binoculars or a telescope. Each object's right ascension and declination are included; these coordinates have been updated for the year 2000. The information in the tables is self-explanatory, for the most part. In the column labeled "Size/Separation/ Period," *size* applies to nonstellar objects, the dimensions of which are given in arc minutes (′). *Separation* indicates the angular distance between stars in double-star systems, in arc seconds (″). *Period* refers to the time over which a variable star's brightness changes from maximum to minimum and back to maximum again.

Stars in multiple-star systems (which are lumped in with double stars in our tables) are designated by the letters A for the brightest star and B for its closest companion, with outlying members labeled C, D, and so on. Separations are given for different pairs of stars in the system, such as AB, AC, or BC.

— Alan M. MacRobert
Associate Editor, *Sky & Telescope*

Northern Hemisphere

A Bull Chock Full

Of the 12 constellations of the zodiac, which one is the most interesting to you? This is admittedly a loaded question, because the word *interesting* is subjective. For example, I find railroads interesting; many commuters find them anything but.

Let's try to answer this question anyway. Scorpius has a striking curved shape formed by relatively bright stars, especially appreciated by our southern readers. Its immediate easterly neighbor, Sagittarius, contains that virtually perfect star picture, the Teapot asterism.

Nevertheless, I'd like to nominate Taurus, the Bull, as my choice for the most interesting zodiacal group, because of its two notable open star clusters, the Pleiades and Hyades. They give the constellation an unquestionably striking appearance, not merely as a collection of stars, but as two stellar concentrations or clumpings. Elsewhere in the sky, the naked-eye cluster Coma Berenices is such an important pattern that it forms a constellation by itself.

The Pleiades and Hyades belong to the deep-sky category of open, or galactic clusters, the latter term recognizing that they are most plentiful along the Milky Way — the plane of our galaxy. However, open clusters are found mainly in the part of the sky where the brightest stars tend to be — and this part dominates the evening heavens in winter, as our chart shows. These clusters and the brightest stars both mark the spiral arms of our galaxy, and in looking at the winter sky we are looking right into the local one, not surprisingly called the Orion arm.

The Pleiades and Hyades are relatively close to us as clusters go, at about 400 and 150 light-years, respectively. Unlike so many other open clusters, however

Interesting Objects

Object	Constellation	Type	Right Ascension	Dec.	Size/Sep./ Period	Magnitude
Hyades	Taurus	OC	$04^h\ 27^m$	+16° 00′	6°	0.5
M1 (Crab Nebula)	Taurus	DN	$05^h\ 34^m.5$	+22° 01′	6′ × 4′	8
M34	Perseus	OC	$02^h\ 02^m.0$	+42° 47′	35′	5.2
M36	Auriga	OC	$05^h\ 36^m.1$	+34° 08′	12′	6.0
M37	Auriga	OC	$05^h\ 52^m.4$	+32° 33′	24′	5.6
M38	Auriga	OC	$05^h\ 28^m.7$	+35° 50′	20′	6.4
M42 (Orion Nebula)	Orion	DN	$05^h\ 35^m.4$	–05° 27′	66′ × 60′	4
M45 (Pleiades)	Taurus	OC	$03^h\ 47^m.0$	+24° 07′	110′	1.2
M81	Ursa Major	G	$09^h\ 55^m.6$	+69° 04′	26′ × 14′	6.8
M82	Ursa Major	G	$09^h\ 55^m.8$	+69° 41′	11′ × 5′	8.4
M103	Cassiopeia	OC	$01^h\ 33^m.2$	+60° 42′	6′	7
NGC 869/884	Perseus	OC	$02^h\ 19^m.0$	+57° 09′	30′	4
(Double Cluster)		OC	$02^h\ 22^m.4$	+57° 07′	30′	4
NGC 1535	Eridanus	PN	$04^h\ 14^m.2$	–12° 44′	0′.7	10
γ Andromedae	Andromeda	DS	$02^h\ 03^m.9$	+42° 20′	AB 9″.8	2.2, 5.1
(Almach)					BC 0″.5	5.5, 6.3
β Persei (Algol)	Perseus	VS	$03^h\ 08^m.2$	+40° 57′	2.9 days	2.1–3.4

DN = diffuse nebula; DS = double or multiple star; DkN = dark nebula; G = galaxy; GC = globular cluster; OC = open cluster; PN = planetary nebula; VS = variable star

January

rich in stars, these two have distinctive shapes. The Pleiades looks like a tiny dipper, which novices frequently think is the Little Dipper, while the Hyades stars form a familiar V. As many readers already know, 1st-magnitude Aldebaran is *not* a member of the cluster but a foreground object only 60 light-years away. By an amazing coincidence, it is placed perfectly along our line of sight to complete the V.

This pattern certainly did not escape the notice of early peoples, and we see below what these stars represented to various cultures around the world.

Taurus is also the home of the well-known Crab Nebula. We know today that it is the remnant of a violent stellar explosion visible in our sky as a supernova in the year 1054. The formation of the Crab Nebula has been a popular topic for planetarium shows. However, some of these presentations, as well as popular writings, state that the Crab Nebula resulted from a supernova explosion in 1054. Really? If that were literally so, we wouldn't have seen it yet. The year 1054 is when the light of the explosion reached Earth; the nebula is some 4,000 light-years away.

M1 in Taurus

The Pleiades in Taurus

In addition to containing the Pleiades, Hyades, and Crab Nebula, Taurus enjoys the distinction of having a star that isn't. This is not unique; misidentifications and mislabelings have left us with scores of "missing" stars.

The object in question is 34 Tauri, which appears in the historic star atlas and catalogue of John Flamsteed, England's first Astronomer Royal. He observed it on December 23, 1690, as a 6th-magnitude star just south of the ecliptic, close to 4^h right ascension. However, 34 Tauri is not exactly missing — it can now be found on the other side of the sky.

You guessed it: it's a planet. In March 1781, William Herschel first recognized it as a solar-system object and called it Georgium Sidus, the Georgian Star, after his monarch, George III of England. Today we call it Uranus.

— George Lovi

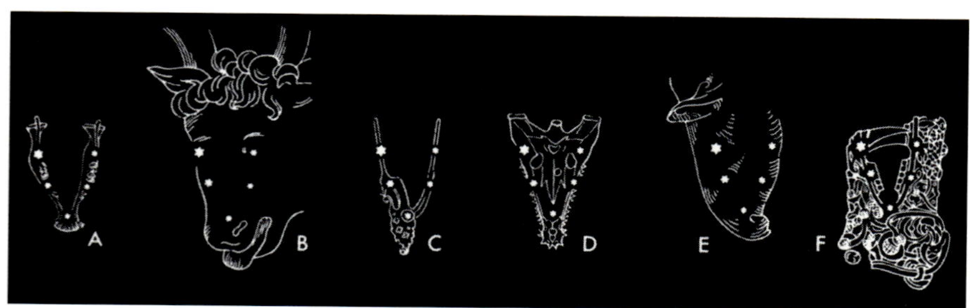

Impressions of the Hyades from different parts of the world: ***A,*** *bull's jaw, ancient Babylonians and present-day Dyaks of Borneo;* ***B,*** *bull's head, classical antiquity;* ***C,*** *betel-nut crackers, Indonesia;* ***D,*** *crocodile's skull, New Guinea;* ***E,*** *tapir's head, South America;* ***F,*** *wolf's head, ancient German tribes. From Helmut Werner,* From the Aratus Globe to the Zeiss Planetarium.

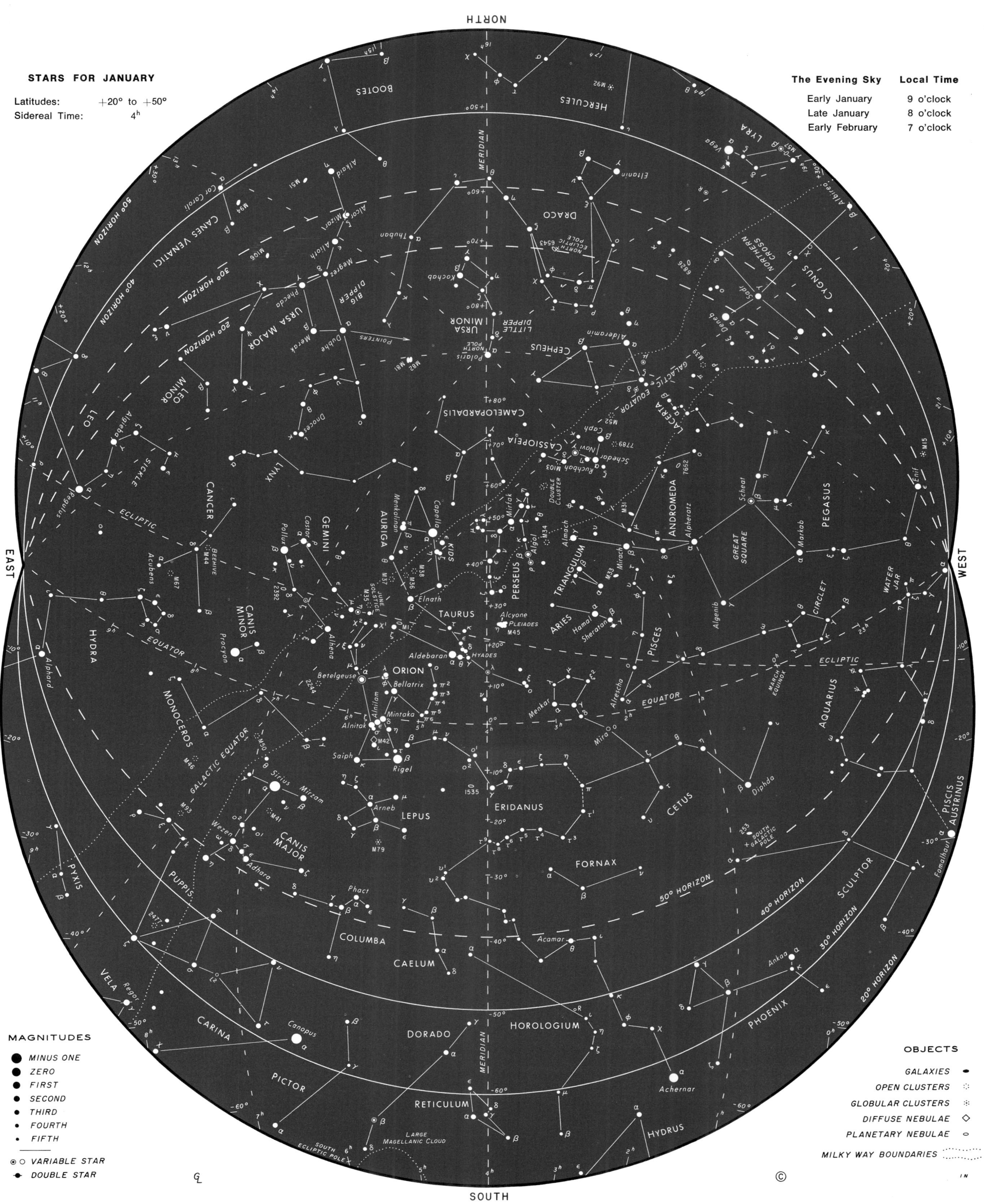
STARS FOR JANUARY
Latitudes: +20° to +50°
Sidereal Time: 4h
The Evening Sky Local Time
Early January 9 o'clock
Late January 8 o'clock
Early February 7 o'clock
NORTH
SOUTH
EAST
WEST
MAGNITUDES
MINUS ONE
ZERO
FIRST
SECOND
THIRD
FOURTH
FIFTH
VARIABLE STAR
DOUBLE STAR
OBJECTS
GALAXIES
OPEN CLUSTERS
GLOBULAR CLUSTERS
DIFFUSE NEBULAE
PLANETARY NEBULAE
MILKY WAY BOUNDARIES
BOOTES
HERCULES
LYRA
DRACO
CYGNUS
NORTHERN CROSS
CANES VENATICI
URSA MAJOR
BIG DIPPER
URSA MINOR
LITTLE DIPPER
CEPHEUS
CAMELOPARDALIS
CASSIOPEIA
LACERTA
LEO MINOR
LEO
SICKLE
LYNX
CANCER
GEMINI
AURIGA
PERSEUS
ANDROMEDA
TRIANGULUM
PEGASUS
GREAT SQUARE
CANIS MINOR
HYDRA
TAURUS
ARIES
PISCES
CIRCLET
WATER JAR
ORION
MONOCEROS
AQUARIUS
CETUS
ERIDANUS
LEPUS
CANIS MAJOR
PISCIS AUSTRINUS
PYXIS
PUPPIS
FORNAX
SCULPTOR
COLUMBA
CAELUM
VELA
CARINA
DORADO
HOROLOGIUM
PHOENIX
PICTOR
RETICULUM
HYDRUS
LARGE MAGELLANIC CLOUD
SOUTH ECLIPTIC POLE
SOUTH GALACTIC POLE
NORTH POLE
NORTH ECLIPTIC POLE
ECLIPTIC
EQUATOR
GALACTIC EQUATOR
MERIDIAN
50° HORIZON
40° HORIZON
30° HORIZON
20° HORIZON
Vega
Polaris
Capella
Aldebaran
Betelgeuse
Rigel
Sirius
Procyon
Pollux
Castor
Regulus
Canopus
Achernar
Fomalhaut
Deneb
Alcyone
PLEIADES
HYADES

FEBRUARY

A Tale of Two Stars

ORION, perhaps the best known constellation, now dominates the evening heavens. Its seven leading stars are bright enough even for city dwellers to see Orion's pattern. But imagine how unimpressive he would be without those two 1st-magnitude "anchor" stars, Betelgeuse and Rigel. On the galactic time scale, their glory days are numbered.

Looking at these two luminaries in Orion is a good way to gain some understanding of the nature of stars. Aside from their brilliance, they form perhaps the best naked-eye color contrast pair. Just shift your eyes from one to the other to see the striking difference in hue. One reason the contrast is so marked is that the eye can only distinguish the colors of bright objects. This is why faint stars all appear white. In ordinary daylight, the retinal receptors that operate best are the cones, which perceive colors. Under very dim conditions, the color-blind but much more sensitive rods take over. We should also note that most descriptions of star colors tend to exaggerate their hues. Betelgeuse is not really a "red" star, but somewhat pale orange-beige. In fact, it's hardly redder than an ordinary incandescent light bulb, both of them having color temperatures of about 3,000° Kelvin (roughly 2,700° C or 4,900° F). See this at night by comparing Betelgeuse with a light from a distant window or porch.

The term *color temperature* can help us understand why stars differ in hue. We can more easily conceptualize this by imagining an electric heating element. As it heats up, it first appears dull and ruddy, then turns red, orange, and eventually yellow-white. Similarly, a star's color represents the wavelengths of light at which it emits the most energy, and this depends on its temperature.

Interesting Objects

Object	Constellation	Type	Right Ascension	Dec.	Size/Sep./ Period	Magnitude
M1 (Crab Nebula)	Taurus	DN	05^h 34^m.5	+22° 01′	6′ × 4′	8
M35	Gemini	OC	06^h 08^m.9	+24° 20′	28′	5.1
M36	Auriga	OC	05^h 36^m.1	+34° 08′	12′	6.0
M37	Auriga	OC	05^h 52^m.4	+32° 33′	24′	5.6
M38	Auriga	OC	05^h 28^m.7	+35° 50′	21′	6.4
M41	Canis Major	OC	06^h 47^m.0	–20° 44′	38′	4.5
M42 (Orion Nebula)	Orion	DN	05^h 35^m.4	–05° 27′	66′ × 60′	4
M50	Monoceros	OC	07^h 03^m.2	–08° 20′	16′	5.9
M79	Lepus	GC	05^h 46^m.7	–24° 33′	9′	8.0
M81	Ursa Major	G	09^h 55^m.6	+69° 04′	26′ × 14′	6.8
M82	Ursa Major	G	09^h 55^m.8	+69° 41′	11′ × 5′	8.4
NGC 2244	Monoceros	OC	06^h 32^m.4	+04° 52′	24′	4.8
NGC 2392 (Eskimo Nebula)	Gemini	PN	07^h 29^m.2	+20° 55′	1′	10
α Geminorum (Castor)	Gemini	DS	07^h 34^m.6	+31° 53′	AB 3″.9	1.9, 2.9
					AC 72″.5	8.8
α Orionis (Betelgeuse)	Orion	VS	05^h 55^m.2	+07° 24′	5.8 years	0.4–1.3

DN = diffuse nebula; DS = double or multiple star; DkN = dark nebula; G = galaxy; GC = globular cluster; OC = open cluster; PN = planetary nebula; VS = variable star

Rigel is a blue supergiant star, one of the rarest breeds in our galaxy. But with their enormous brilliance up to 100,000 times that of the Sun and more these stars remain conspicuous over great distances and, for us on Earth, make the winter skies so splendid through constellations like Orion. They also help make the arms of spiral galaxies stand out.

Rigel is even brighter than those blue stars at the top of the main sequence of stable mid-life stars, which run from the rare blue to the increasingly cooler, dimmer, more numerous white, yellow, orange, and red dwarf stars. Red dwarfs are the most numerous stars by far, the "ordinary citizens" of our galaxy, but because of their low luminosity not one is visible to the unaided eye. Main-sequence stars constitute some 90 percent of the stellar population, and all produce their energy by fusing hydrogen into helium deep within their cores. The brighter and hotter a star is, the faster this occurs. In a sense, red dwarfs and bright blue stars could be thought of as, respectively, the candles and the blowtorches of the main sequence — even the colors are analogous.

M37 in Auriga

M42 in Orion

When a main-sequence star accumulates sufficient helium in its core, its energy output increases significantly, and it swells into a red giant or supergiant like Betelgeuse. This is what Rigel will become in a few million years. The Sun has another six billion years to go, an indication of how much more slowly it is consuming its hydrogen.

Red supergiants like Betelgeuse are gigantic bloated globes of distended, cooler gas. If such a star were to replace the Sun in the solar system, it might extend beyond Mars' orbit. Despite the cooler color temperature of a red supergiant, its surface area is enormous enough to make the star intrinsically very bright. Stars like Rigel are smaller, more concentrated light sources.

In a huge star, the nuclear furnace of the core produces successively heavier elements to balance the incessant crush of gravity. From simple beginnings, the fusing of hydrogen into helium, the core manufactures carbon, neon, oxygen, sulfur, and finally iron in thin, onion-layered shells. But once the core begins creating iron, the heaviest element a star can generate through ordinary atomic processes, its days are numbered. The formation of elements heavier than iron consumes rather than produces energy. Having lost its heat source, the core can no longer support the star's vast weight. The core collapses, triggering a cataclysmic supernova explosion. The blast generates elements heavier than iron and scatters them into space.

In a sense, those celestial jewels Betelgeuse and Rigel will become jewelry factories by creating such metals as silver, gold, and platinum. This will be the fate of the other Orion luminaries as well. So if the proper motions of the stars don't disperse this splendid constellation first, the "popping off" of its members will.

— George Lovi

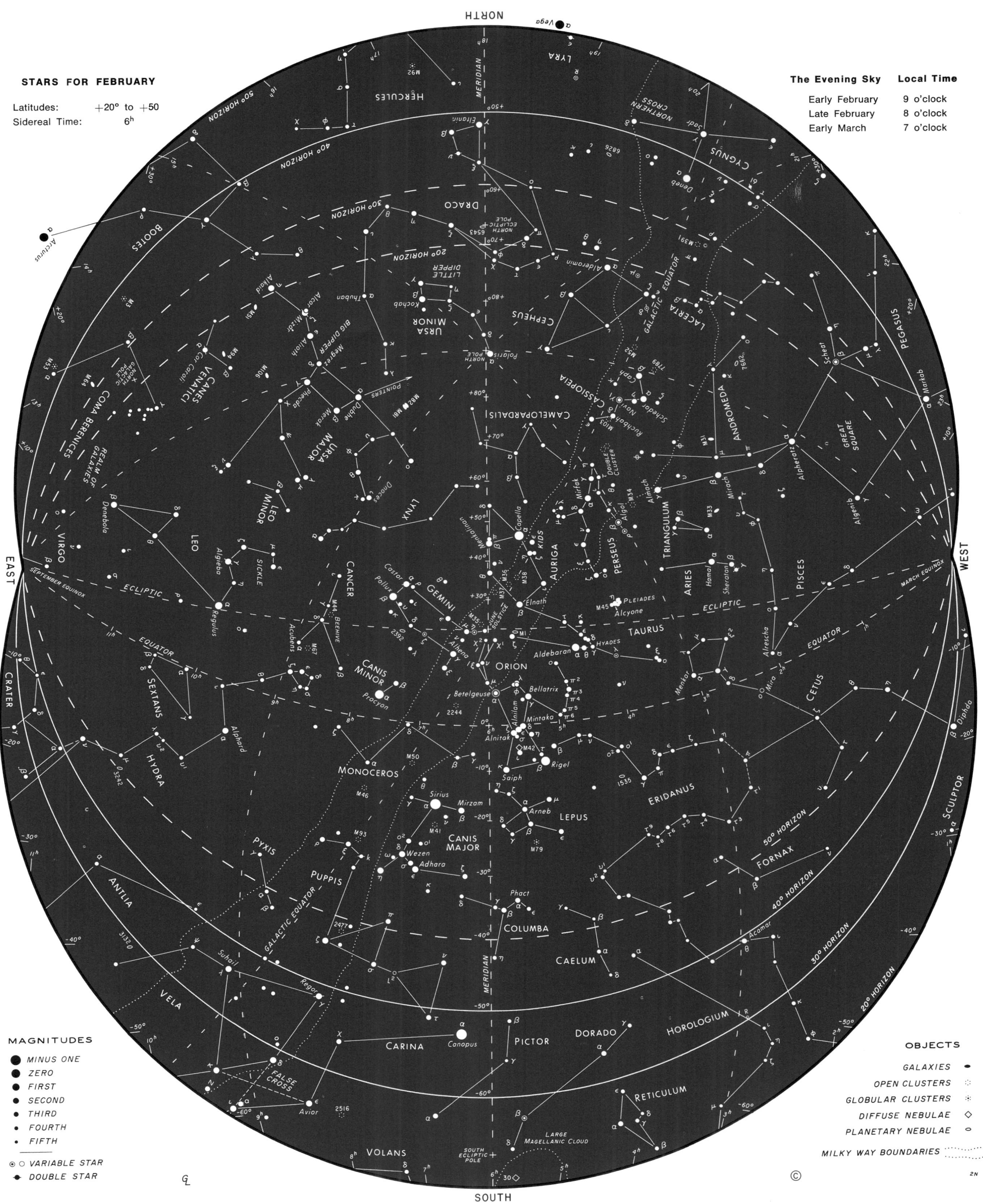
STARS FOR FEBRUARY
Latitudes: +20° to +50
Sidereal Time: 6h
The Evening Sky | Local Time
Early February | 9 o'clock
Late February | 8 o'clock
Early March | 7 o'clock
NORTH
SOUTH
EAST
WEST
MERIDIAN
ECLIPTIC
EQUATOR
GALACTIC EQUATOR
SEPTEMBER EQUINOX
MARCH EQUINOX
NORTH POLE
NORTH ECLIPTIC POLE
SOUTH ECLIPTIC POLE
50° HORIZON
40° HORIZON
30° HORIZON
20° HORIZON
HERCULES
LYRA
Vega
NORTHERN CROSS
CYGNUS
Deneb
BOOTES
Arcturus
DRACO
LITTLE DIPPER
URSA MINOR
Polaris
Thuban
Kochab
CEPHEUS
Alderamin
LACERTA
PEGASUS
Scheat
Markab
GREAT SQUARE
Algenib
ANDROMEDA
Alpheratz
Mirach
Almach
CASSIOPEIA
Caph
Schedar
Navi
Ruchbah
CAMELOPARDALIS
BIG DIPPER
URSA MAJOR
Alcor
Mizar
Alioth
Megrez
Phecda
Merak
Dubhe
POINTERS
CANES VENATICI
Cor Caroli
COMA BERENICES
NORTH GALACTIC POLE
REALM OF GALAXIES
LEO MINOR
LYNX
LEO
Denebola
Regulus
Algieba
SICKLE
VIRGO
CANCER
M44 BEEHIVE
Acubens
GEMINI
Castor
Pollux
Alhena
SUMMER SOLSTICE
AURIGA
Capella
Menkalinan
KIDS
Elnath
PERSEUS
Mirfak
Algol
DOUBLE CLUSTER
TRIANGULUM
ARIES
Hamal
Sheratan
PISCES
CETUS
Menkar
Mira
Diphda
TAURUS
PLEIADES
Alcyone
HYADES
Aldebaran
ORION
Betelgeuse
Bellatrix
Mintaka
Alnilam
Alnitak
Rigel
Saiph
CANIS MINOR
Procyon
MONOCEROS
HYDRA
Alphard
SEXTANS
CRATER
CANIS MAJOR
Sirius
Mirzam
Wezen
Adhara
LEPUS
Arneb
ERIDANUS
FORNAX
SCULPTOR
PYXIS
PUPPIS
ANTLIA
VELA
Suhail
Regor
COLUMBA
Phact
CAELUM
Acamar
CARINA
Canopus
PICTOR
DORADO
HOROLOGIUM
RETICULUM
FALSE CROSS
Avior
VOLANS
LARGE MAGELLANIC CLOUD
MAGNITUDES
MINUS ONE
ZERO
FIRST
SECOND
THIRD
FOURTH
FIFTH
VARIABLE STAR
DOUBLE STAR
OBJECTS
GALAXIES
OPEN CLUSTERS
GLOBULAR CLUSTERS
DIFFUSE NEBULAE
PLANETARY NEBULAE
MILKY WAY BOUNDARIES
©

March

The Heavenly Picture Book

HAVE you ever wondered why particular constellations are formed from the stars in this or that part of the sky? This question is an interesting one to ponder on March evenings, when a good selection of star pictures is spread across the firmament. The bright winter constellations, sinking down in the western sky, will soon leave us. Yet, they are still well placed at chart time (given in the upper right-hand corner of the facing page), so this is a good opportunity to view the winter pageant.

Moving into prominent position near the meridian are the constellations of spring. While they are, for the most part, not as brilliant as the winter group, they are no less important. The entourage contains such notables as Leo, the Lion, and Ursa Major, the Larger Bear.

In a number of languages the word for constellation literally means star picture — for example, the German *sternbild.* Although some authors would have us believe that constellations represent what they do because the stars form an easily perceived pattern of the object, this is usually far from the truth. There are exceptions, however — Orion does suggest the standing figure of a person, and Ursa Major resembles a stick-figure rendition of a four-legged creature, though not necessarily a bear. Why then have the constellations become what they are?

It is fairly certain that the oldest star patterns recognized today date from the earliest Mideastern peoples, particularly those, like the Sumerians, who inhabited the Tigris-Euphrates region. They used pictures of constellations, such as Leo, on their pottery, tablets, and cylinder seals.

Perhaps most influential in "officializing" today's roster of ancient constellations was the epic poem *Phaenomena* by the noted Greek Aratus (also spelled Aratos) of Soli, who lived from about 315 to 245 B.C. His writings mention some 45 star groups that represent somewhat of a synthesis of those noted by earlier authors. Among these writers was Eudoxus of Cnidus, who lived in the previous century.

Was there some sort of master plan to the constellations? This question was considered by Helmut Werner in his article "The Composition of the Classical Constellation Sky According to Primary Themes" in the October 1961 issue of *Planetarium News.* Werner, a leading authority on constellation lore spanning a wide variety of cultures, marshals strong evidence for a "systems approach" to the heavenly picture book. According to him, one of its main themes is human mastery over the animal world.

> We notice that the animals are by far the more numerous. They are arranged in a certain way with respect to different constellations representing humans, that is to say in accordance with the theme of mastery over the animals. Relevant verses in the didactic poem on Astronomy of *Aratus* point this out clearly several times. . . . Thus, according to *Aratus* and his later followers, Hercules is associated with the Dragon, the Serpent-bearer with the Scorpion, the Charioteer with the Bull, the Water-carrier with the Goat-fish, the Archer with the Eagle, Orion with the Great Dog and the Hare, and the Centaur with the Wolf.

Werner has proposed that Hercules has a leading role in the heavenly depiction of our rule over the animals, since this ancient strong man wears the skin of the ferocious and deadly Nemean Lion, which he slew. Furthermore, he is standing on the head of another vicious beast, Draco, the Dragon. On our chart, Hercules's feet are in the process of rising above the northeast horizon with the Dragon underfoot. (Incidentally, today Hercules seems to be upside down, but in ancient times the sky's precessional shift would have made him appear upright in the north.)

M35 in Gemini

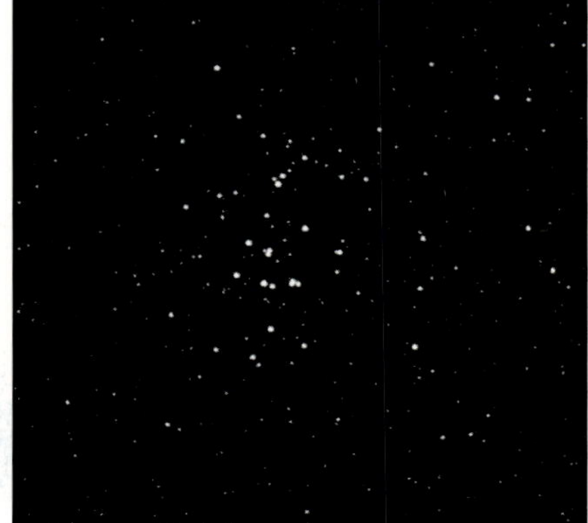

M44 in Cancer

In the west, among the luminous winter contingent is the charioteer Auriga holding a goat (Capella) and three kids in his arm. Also significant, according to Werner, is Auriga's position atop the horns of Taurus, the Bull, which he regards as an act of domination.

Preceding the Bull along the zodiac is Aries, the Ram, which is usually pictured with its head facing Taurus and Auriga. This, Werner maintains, indicates an association between these constellations, especially when we consider that sheep, goats, and cattle are among the earliest domestic animals.

As for Aries, I wonder if its placement might have a biblical implication as well. At the time of year when the Sun is in that part of the sky, two major festivals — Easter for Christians and Passover for Jews — are celebrated. In both holidays the Paschal Lamb (the same species as Aries) plays a central role.

There are also groups of constellations that are related mythologically. The best example is the royal family of the sky: Perseus, Andromeda, Cassiopeia, and Cepheus. They are all grouped together and are visible this month in the west-northwest. Two creatures associated with them — Pegasus, the Winged Horse, and Cetus, the Whale — are also nearby. The former, however, has already set by chart time in March, and the latter's head is about to sink out of sight.

— George Lovi

Interesting Objects

Object	Constellation	Type	Right Ascension	Dec.	Size/Sep./ Period	Magnitude
M35	Gemini	OC	$06^h\ 08^m.9$	+24° 20′	28′	5.1
M36	Auriga	OC	$05^h\ 36^m.1$	+34° 08′	12′	6.0
M37	Auriga	OC	$05^h\ 52^m.4$	+32° 33′	24′	5.6
M38	Auriga	OC	$05^h\ 28^m.7$	+35° 50′	21′	6.4
M44 (Beehive Cluster)	Cancer	OC	$08^h\ 40^m.1$	+19° 59′	95′	3.1
M46	Puppis	OC	$07^h\ 41^m.8$	–14° 49′	27′	6.1
M50	Monoceros	OC	$07^h\ 03^m.2$	–08° 20′	16′	5.9
M67	Cancer	OC	$08^h\ 50^m.4$	+11° 49′	30′	6.9
M81	Ursa Major	G	$09^h\ 55^m.6$	+69° 04′	26′ × 14′	6.8
M82	Ursa Major	G	$09^h\ 55^m.8$	+69° 41′	11′ × 5′	8.4
M93	Puppis	OC	$07^h\ 44^m.6$	–23° 52′	22′	6
NGC 2392 (Eskimo Nebula)	Gemini	PN	$07^h\ 29^m.2$	+20° 55′	1′	10
α Geminorum (Castor)	Gemini	DS	$07^h\ 34^m.6$	+31° 53′	AB 3″.9 AC 72″.5	1.9, 2.9 8.8
γ Leonis (Algeiba)	Leo	DS	$10^h\ 20^m.0$	+19° 51′	4″.4	2.2, 3.5

DN = diffuse nebula; DS = double or multiple star; DkN = dark nebula; G = galaxy; GC = globular cluster; OC = open cluster; PN = planetary nebula; VS = variable star

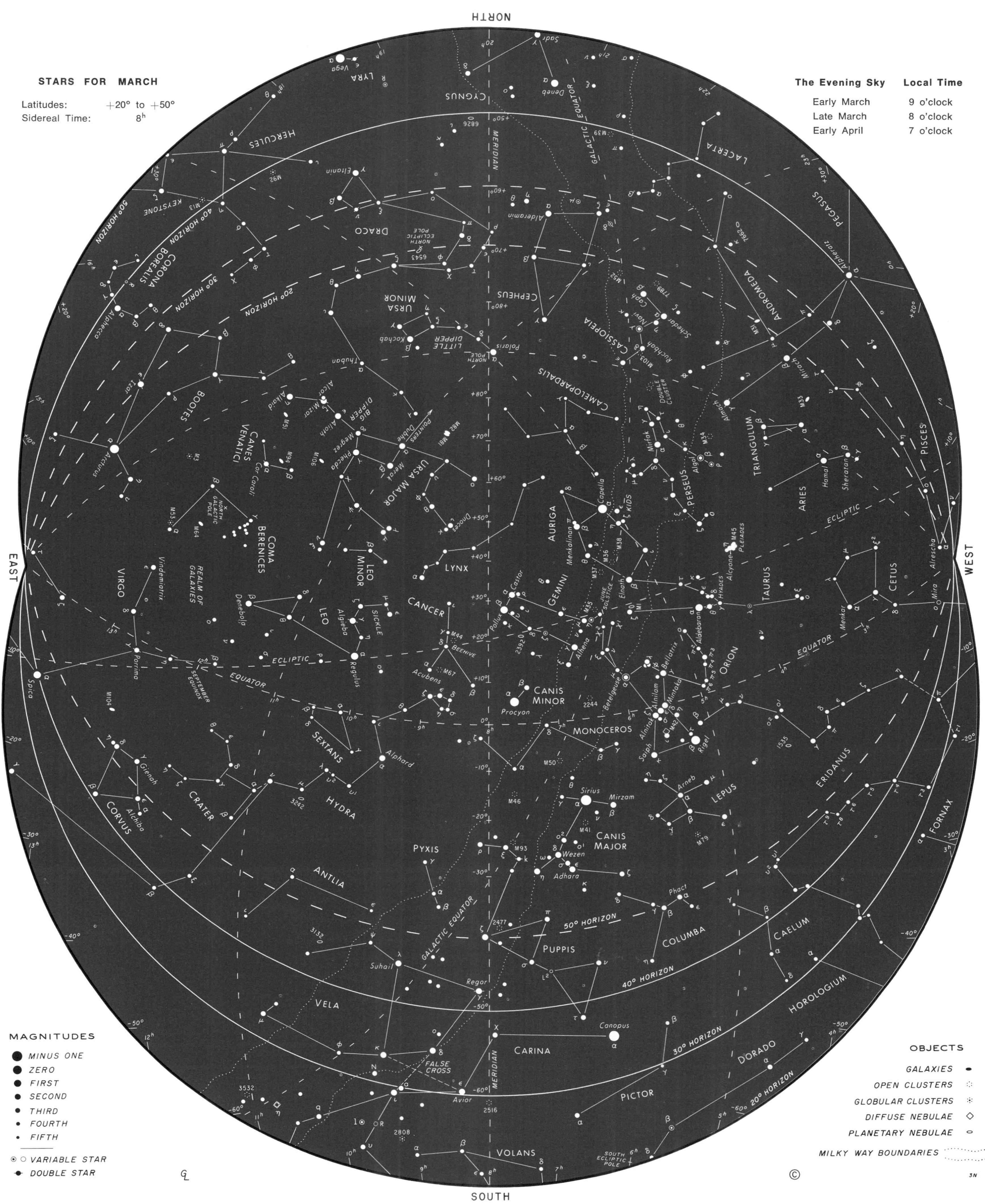

STARS FOR MARCH
Latitudes: +20° to +50°
Sidereal Time: 8h
The Evening Sky Local Time
Early March 9 o'clock
Late March 8 o'clock
Early April 7 o'clock
NORTH
SOUTH
EAST
WEST
MERIDIAN
ECLIPTIC
EQUATOR
GALACTIC EQUATOR
50° HORIZON
40° HORIZON
30° HORIZON
20° HORIZON
LYRA
CYGNUS
HERCULES
LACERTA
PEGASUS
DRACO
CORONA BOREALIS
URSA MINOR
LITTLE DIPPER
CEPHEUS
ANDROMEDA
CASSIOPEIA
CAMELOPARDALIS
BOOTES
CANES VENATICI
BIG DIPPER
URSA MAJOR
TRIANGULUM
PERSEUS
ARIES
PISCES
COMA BERENICES
AURIGA
LEO MINOR
LYNX
VIRGO
REALM OF GALAXIES
GEMINI
TAURUS
CETUS
CANCER
LEO
SICKLE
ORION
CANIS MINOR
MONOCEROS
SEXTANS
HYDRA
CRATER
CORVUS
ERIDANUS
LEPUS
CANIS MAJOR
PYXIS
ANTLIA
FORNAX
PUPPIS
COLUMBA
CAELUM
VELA
HOROLOGIUM
CARINA
FALSE CROSS
DORADO
PICTOR
VOLANS
NORTH ECLIPTIC POLE
NORTH POLE
NORTH GALACTIC POLE
SOUTH ECLIPTIC POLE
SEPTEMBER EQUINOX
JUNE SOLSTICE
PLEIADES
HYADES
BEEHIVE
KIDS
KEYSTONE
POINTERS
MAGNITUDES
MINUS ONE
ZERO
FIRST
SECOND
THIRD
FOURTH
FIFTH
VARIABLE STAR
DOUBLE STAR
OBJECTS
GALAXIES
OPEN CLUSTERS
GLOBULAR CLUSTERS
DIFFUSE NEBULAE
PLANETARY NEBULAE
MILKY WAY BOUNDARIES

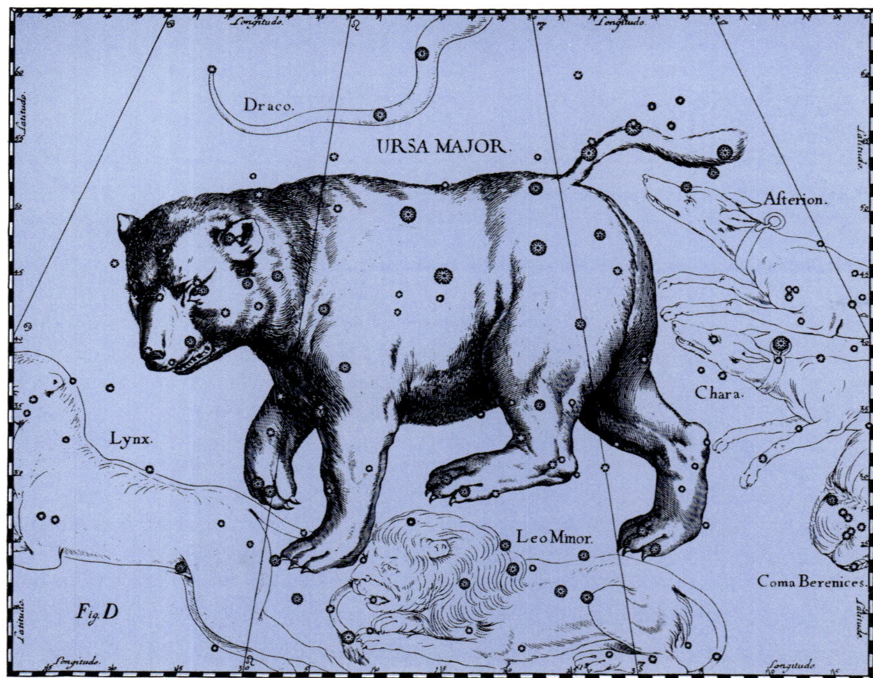

April

Anatomy of a Great Bear

ON balmy April evenings we must bid farewell to the glittering winter stars, still visible briefly in the western heavens. This scene has for me since childhood held a touch of sadness — as when we say so long for a while to good friends, even though we know they'll be back. These friends will return to the evening sky late this year, but readers who cannot wait should set their alarms for the predawn hours after mid-August.

As if to compensate for the departure of our winter luminaries, nature has provided a striking and well-known pattern high in the north at chart time: that seven-star asterism we in America call the Big Dipper and many other cultures regard as some type of carriage or wheeled vehicle (to the British, it's the Plough). Regardless of what these seven stars portray to various peoples, to Western civilization they have always been part of the larger constellation of the Great Bear, which we call Ursa Major. The stars immediately to the west and south of our Dipper form a pretty good stick figure of this creature.

Both Orion and the Big Dipper are prominent, easy-to-remember patterns. In both instances the patterns are not completely the result of chance lineups of stars along our line of sight, as is so often the case with constellations. Orion's principal stars (except Betelgeuse) belong to one of those loose stellar groupings called *associations.* The Big Dipper's stars, with the exception of the end ones, Dubhe and Alkaid, are also part of a loose cluster, although not a classic association. My favorite 19th-century astronomy popularizer, Richard A. Proctor, back in 1869 first called attention to the common proper motions shared by most of the Dipper's stars. Later work confirmed that these stars indeed are moving together through space on parallel paths, along with some fainter neighbors, as shown in the diagram on this page.

Many bright stars in various parts of the sky have nearly the same motion as the Ursa Major cluster, and form the so-called Ursa Major stream. One star listed as a probable member of the stream is Sirius in Canis Major; another is Alphecca in Corona Borealis (low in the east on this month's chart).

What is really remarkable about the Big Dipper stars is that in one way or another they constituted a bear to widely separated early peoples — not only Old World ancients but New World Native American tribes as well. The bowl of the Dipper was a bear to certain St. Lawrence basin tribes, with the handle stars being in pursuit.

But *our* bear, as indicated, is considerably larger. Its nose is 3rd-magnitude Omicron, whose name Muscida, according to star-name expert George A. Davis Jr. is a corruption of the Latin *musus* or *musum,* meaning mouth or muzzle.

The paws of the beast are marked by an almost equally spaced and oriented set of three naked-eye stellar pairs, which are even lined up nearly perfectly. They were known to the early Arabs as the "Three Leaps of the Gazelle," from which their names derive. The stars Iota (ι) and Kappa (κ), in the front paws, represent the third leap. Iota has the traditional name Talitha, but on our chart we label it Dnoces, after the NASA-inspired name, which is nothing more than the word *second* spelled backward. Our space agency decided to commemorate the three astronauts who perished in the Apollo 1 fire of January 1967, and named this star after one of them, Edward H. White II.

M81 in Ursa Major

Since World War II a number of non-traditional fabricated names have been assigned to standard navigation stars that did not already have names (such as Menkent for Theta Centauri, low in the southeast on this month's chart). However, some people — including me — have objected to a fabrication such as Dnoces being allowed to usurp an existing traditional star name. Why couldn't Dnoces have been applied, say, to Theta Ursae Majoris?

The next "leap" pair of stars forms one of the hind paws of the Bear. These stars, Lambda (λ) and Mu (μ), have the names Tania Borealis and Tania Australis, respectively, meaning "the northern (star) of the second leap" and "the southern (star) of the second leap." These names derive from both Arabic and Latin. The easternmost "leap," Nu (ν) and Xi (ξ) (in the other hind paw), are Alula Borealis and Alula Australis, which refer similarly to the first leap. Alula Australis is a famous visual binary star that enjoys the distinction of being the first one whose orbit was calculated, back in 1828.

In pointing out the Great Bear to lay people, one often has to explain its long tail. Perhaps the most common reason given is that attributed to the 16th-century British writer Thomas Hood: "Imagine that Jupiter, fearing to come too nigh unto her teeth, layde holde on her tayle, and thereby drewe her up into the heaven; so that shee of herself being very weightie, and the distance from the earth to the heavens very great, there was great likelihood that her tayle must stretch. Other reason know I none."

— *George Lovi*

Interesting Objects

Object	Constellation	Type	Right Ascension	Dec.	Size/Sep./ Period	Magnitude
Coma Cluster	Coma Ber.	OC	$12^{h}\ 25^{m}$	+26° 00′	4°.5	1.8
M44 (Beehive Cluster)	Cancer	OC	$08^{h}\ 40^{m}.1$	+19° 59′	95′	3.1
M67	Cancer	OC	$08^{h}\ 50^{m}.4$	+11° 49′	30′	6.9
M81	Ursa Major	G	$09^{h}\ 55^{m}.6$	+69° 04′	26′ × 14′	6.8
M82	Ursa Major	G	$09^{h}\ 55^{m}.8$	+69° 41′	11′ × 5′	8.4
M94	Canes Ven.	OC	$12^{h}\ 50^{m}.9$	+41° 07′	11′ × 9′	8.1
M106	Canes Ven.	G	$12^{h}\ 19^{m}.0$	+47° 18′	18′ × 8′	8.3
NGC 3242 (Ghost of Jupiter)	Hydra	PN	$10^{h}\ 24^{m}.8$	−18° 38′	21′	9
NGC 6543	Draco	PN	$17^{h}\ 58^{m}.6$	+66° 38′	5′.8	9
α Canum Venaticorum	Canes Ven.	DS	$12^{h}\ 56^{m}.0$	+38° 19′	19″.4	2.9, 5.5
γ Leonis (Algeiba)	Leo	DS	$10^{h}\ 20^{m}.0$	+19° 51′	4″.4	2.2, 3.5
ζ Ursae Majoris (Mizar)	Ursa Major	DS	$13^{h}\ 23^{m}.9$	+54° 56′	AB 14″.4 AC 709″	2.3, 4.0 4.0

DN = diffuse nebula; DS = double or multiple star; DkN = dark nebula; G = galaxy; GC = globular cluster; OC = open cluster; PN = planetary nebula; VS = variable star

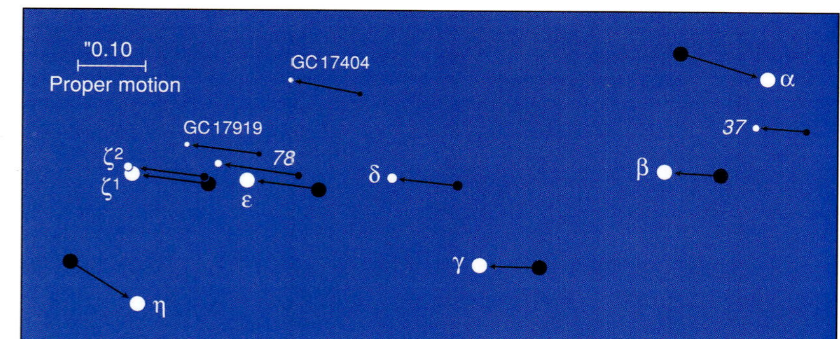

The Big Dipper's stars are moving together through space, with their current positions shown in black and those of 100,000 years from now in white. The end stars Alkaid (η) and Dubhe (α) are not cluster members.

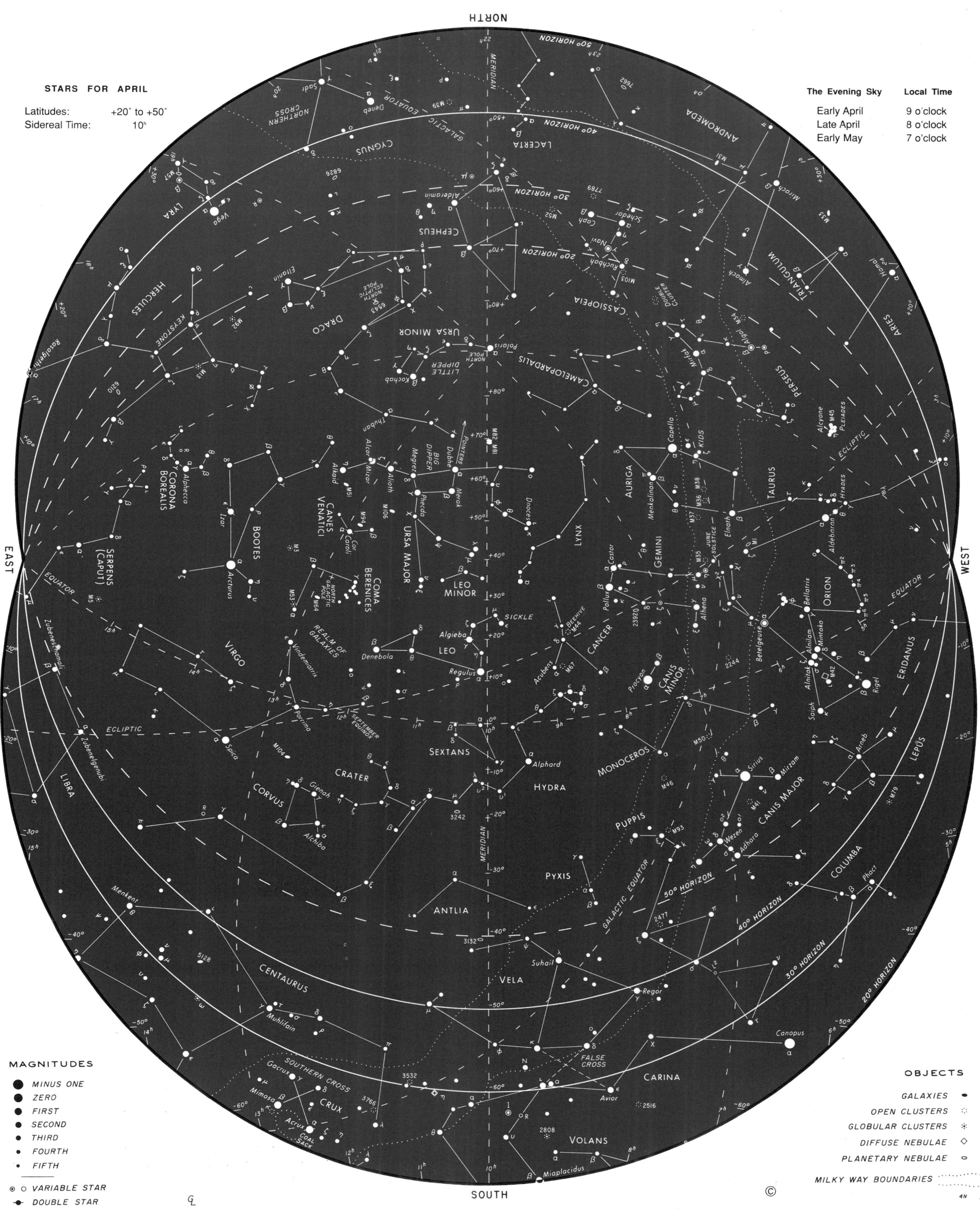
STARS FOR APRIL
Latitudes: +20° to +50°
Sidereal Time: 10h
The Evening Sky / Local Time
Early April 9 o'clock
Late April 8 o'clock
Early May 7 o'clock
NORTH
SOUTH
EAST
WEST
MERIDIAN
ECLIPTIC
EQUATOR
GALACTIC EQUATOR
50° HORIZON
40° HORIZON
30° HORIZON
20° HORIZON
CYGNUS
LACERTA
ANDROMEDA
LYRA
CEPHEUS
CASSIOPEIA
TRIANGULUM
HERCULES
KEYSTONE
DRACO
URSA MINOR
LITTLE DIPPER
CAMELOPARDALIS
PERSEUS
ARIES
CORONA BOREALIS
BOOTES
CANES VENATICI
BIG DIPPER
URSA MAJOR
LYNX
AURIGA
TAURUS
SERPENS (CAPUT)
COMA BERENICES
LEO MINOR
GEMINI
ORION
SICKLE
LEO
CANCER
VIRGO
REALM OF GALAXIES
CANIS MINOR
ERIDANUS
SEXTANS
MONOCEROS
LEPUS
LIBRA
CRATER
CORVUS
HYDRA
CANIS MAJOR
PUPPIS
PYXIS
COLUMBA
ANTLIA
CENTAURUS
VELA
CARINA
SOUTHERN CROSS
CRUX
COAL SACK
FALSE CROSS
VOLANS
MAGNITUDES
MINUS ONE
ZERO
FIRST
SECOND
THIRD
FOURTH
FIFTH
VARIABLE STAR
DOUBLE STAR
OBJECTS
GALAXIES
OPEN CLUSTERS
GLOBULAR CLUSTERS
DIFFUSE NEBULAE
PLANETARY NEBULAE
MILKY WAY BOUNDARIES
©

MAY

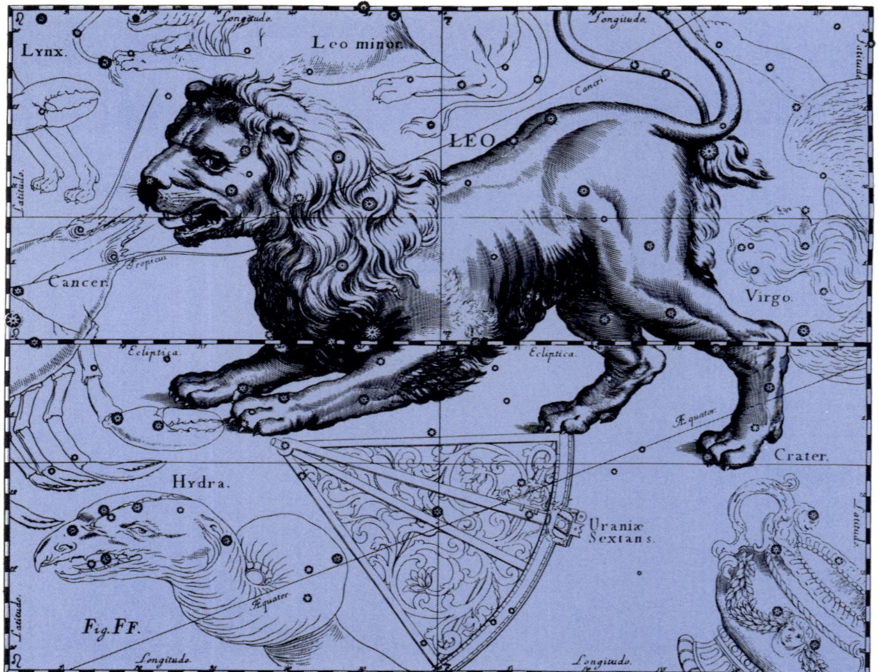

Cosmic Clusters Large and Small

STARS, like people, tend to be gregarious and enjoy the company of others. Single stars like the Sun are in the minority. Astronomers have found that most stars are members of double or multiple systems. There are also larger groupings such as stellar associations, clusters, and galaxies. All this, of course, is due to gravitation.

Galaxies also like to congregate, and this month the most prominent example of a galactic get-together is high in the south just east of the meridian between the main stars of Leo and Virgo. Labeled "Realm of the Galaxies" on the chart, this cluster is also known as the Virgo Cloud.

Multiple-star systems are scattered across the firmament, and, because of color contrasts between component stars, many of these systems are beautiful in telescopes. One of the best-known multiples is right at the bend in the Big Dipper's handle, high in the north. Mizar and Alcor are the classic naked-eye pair often cited as a test for good eyesight. This is an optical pair only, meaning that the stars appear together along our line of sight, but they are not actually close together in space.

However, Mizar itself is a true double star, the first one ever discovered, back in 1650 by Giambattista Riccioli. These stars are actually a physical pair that orbit about a common center of gravity, called a *barycenter*. The distance of each star from this common point is related to the star's mass. Indeed, careful observation of such binary systems can provide information on the orbits and, in turn, the stars' masses.

Some other leading doubles are visible during May evenings. Epsilon (ε) Boötis (Izar), Gamma (γ) Virginis (Porrima), and Gamma Leonis (Algeiba) are well worth observing. The first is a prime example of a beautiful color contrast,

Interesting Objects

Object	Constellation	Type	Right Ascension	Dec.	Size/Sep./ Period	Magnitude
Coma Cluster	Coma Ber.	OC	$12^h\ 25^m$	+26° 00′	4°.5	1.8
M3	Canes Ven.	GC	$13^h\ 42^m.2$	+28° 23′	16′	6.4
M51 (Whirlpool Galaxy)	Canes Ven.	G	$13^h\ 29^m.9$	+47° 12′	11′ × 8′	8.1
M64 (Black-eye Galaxy)	Coma Ber.	G	$12^h\ 56^m.7$	+21° 41′	9′ × 5′	8.5
M81	Ursa Major	G	$09^h\ 55^m.6$	+69° 04′	26′ × 14′	6.8
M82	Ursa Major	G	$09^h\ 55^m.8$	+69° 41′	11′ × 5′	8.4
M94	Canes Ven.	OC	$12^h\ 50^m.9$	+41° 07′	11′ × 9′	8.1
M104 (Sombrero Gal.)	Virgo	G	$12^h\ 40^m.0$	−11° 37′	9′ × 4′	8.3
M106	Canes Ven.	G	$12^h\ 19^m.0$	+47° 18′	18′ × 8′	8.3
NGC 3242 (Ghost of Jupiter)	Hydra	PN	$10^h\ 24^m.8$	−18° 38′	21′	9
ε Boötis (Izar)	Boötes	DS	$14^h\ 45^m.0$	+27° 04′	2″.8	2.5, 4.9
α Canum Venaticorum	Canes Ven.	DS	$12^h\ 56^m.0$	+38° 19′	19″.4	2.9, 5.5
γ Leonis (Algeiba)	Leo	DS	$10^h\ 20^m.0$	+19° 51′	4″.4	2.2, 3.5
ζ Ursae Majoris (Mizar)	Ursa Major	DS	$13^h\ 23^m.9$	+54° 56′	AB 14″.4 AC 709″	2.3, 4.0 4.0

DN = diffuse nebula; DS = double or multiple star; DkN = dark nebula; G = galaxy; GC = globular cluster; OC = open cluster; PN = planetary nebula; VS = variable star

so much so that Wilhelm Struve, now recognized as the father of double-star astronomy, gave it the Latinized nickname *Pulcherrima*.

Moving up in complexity from double stars, we come to a type of stellar aggregation only recognized in this century: an association. Associations are composed of hot, young, blue stars that were born out of the same cloud of gaseous material and are still near one another in space. The main stars of Orion constitute such a grouping, while the more extensive Scorpius-Centaurus Association rises in the southeast on the facing chart.

Next are the open or galactic star clusters that, in a sense, are more populous and tightly grouped associations. They too consist of young stars. The best example in our sky is the beautiful naked-eye Pleiades Cluster, which, alas, has just set at chart time. The constellation Coma Berenices, close to the meridian this month, is based on another galactic cluster.

As I mentioned in my essay accompanying the January star chart, open clusters follow the Milky Way band, indicating their position in space near the galactic plane. The two associations cited above also lie within the galaxy's disk.

M3 in Canes Venatici

M51 in Canes Venatici

The other major type of star cluster, much more densely populated than open clusters, is the globular. Such clusters lie not in the galactic plane but instead form a spherical halo around the Milky Way's center, and, unlike the more numerous open clusters, contain some of the oldest stars in our galaxy.

Just above the 40° horizon in the south, east of the meridian, is the most prominent globular cluster in our heavens, Omega (ω) Centauri. East of the zenith in Canes Venatici is M3, pictured above, which is better known by amateurs because of its favorable sky location for Northern Hemisphere observers.

Galaxies themselves make up the next class of stellar conglomerations, and they come in a number of varieties. Yet there is no sharp line of demarcation between clusters and galaxies. For example, a tiny spherical galaxy can be similar to a diffuse globular cluster in structure and composition. Indeed, spherical and elliptical galaxies can sometimes be found in the outer halo of a large spiral system. The Andromeda Galaxy has two such companions.

Finally, there is togetherness on an intergalactic scale. The Milky Way is part of a loose "association" called the Local Group of galaxies. Its nearly 20 members include the Magellanic Clouds and the Andromeda and Triangulum spiral galaxies (M31 and M33).

Today we have identified many clusters of galaxies. Indeed, it seems that most galaxies come in clusters or at least small groups. In Corona Borealis alone nearly 100 clusters of galaxies have been identified.

Over the past few decades astronomers have determined that the Milky Way is part of an enormous assemblage of galaxies. The plane of this "supercluster" lies virtually perpendicular to that of our home system, running through the Virgo Cloud and extending past the Big Dipper, then toward Cassiopeia, Andromeda, Pisces, Sculptor, Tucana, onward past the south polar area, and back up through Circinus-Centaurus to Virgo. One might say the Virgo Cloud is the "Sagittarius of the Supercluster" — that is, its center.

— George Lovi

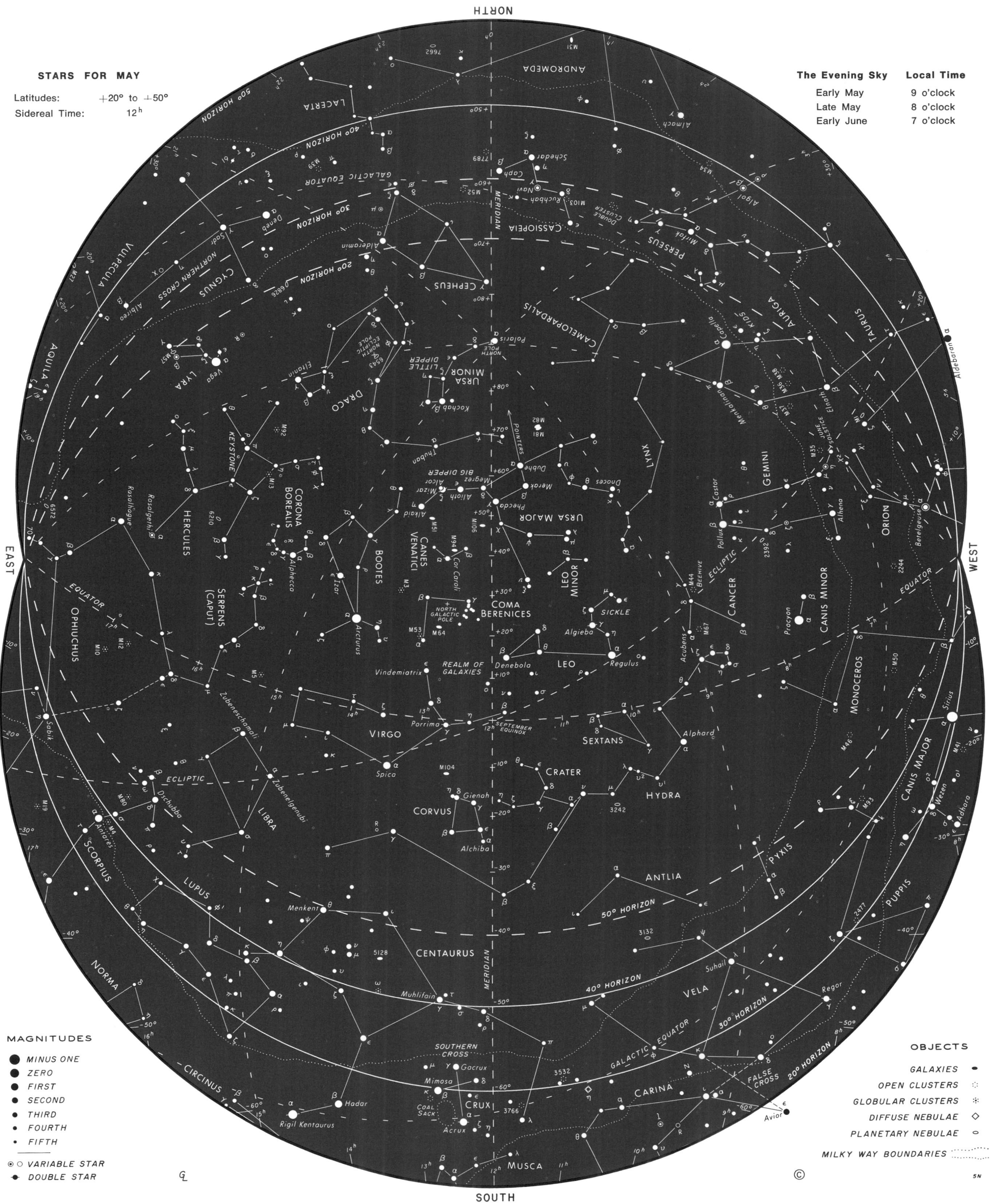
STARS FOR MAY
Latitudes: +20° to +50°
Sidereal Time: 12h
The Evening Sky Local Time
Early May 9 o'clock
Late May 8 o'clock
Early June 7 o'clock
NORTH
SOUTH
EAST
WEST
MAGNITUDES
MINUS ONE
ZERO
FIRST
SECOND
THIRD
FOURTH
FIFTH
VARIABLE STAR
DOUBLE STAR
OBJECTS
GALAXIES
OPEN CLUSTERS
GLOBULAR CLUSTERS
DIFFUSE NEBULAE
PLANETARY NEBULAE
MILKY WAY BOUNDARIES
ANDROMEDA
LACERTA
CASSIOPEIA
CEPHEUS
PERSEUS
AURIGA
TAURUS
CYGNUS
NORTHERN CROSS
VULPECULA
AQUILA
LYRA
DRACO
URSA MINOR
LITTLE DIPPER
CAMELOPARDALIS
LYNX
GEMINI
ORION
HERCULES
KEYSTONE
CORONA BOREALIS
BOOTES
CANES VENATICI
URSA MAJOR
BIG DIPPER
POINTERS
LEO MINOR
CANCER
CANIS MINOR
OPHIUCHUS
SERPENS (CAPUT)
COMA BERENICES
NORTH GALACTIC POLE
SICKLE
LEO
REALM OF GALAXIES
MONOCEROS
VIRGO
SEPTEMBER EQUINOX
SEXTANS
HYDRA
CRATER
CORVUS
LIBRA
SCORPIUS
CANIS MAJOR
PUPPIS
PYXIS
ANTLIA
LUPUS
CENTAURUS
NORMA
VELA
SOUTHERN CROSS
CRUX
COAL SACK
CIRCINUS
CARINA
FALSE CROSS
MUSCA
ECLIPTIC
EQUATOR
GALACTIC EQUATOR
MERIDIAN
20° HORIZON
30° HORIZON
40° HORIZON
50° HORIZON
Polaris
Vega
Deneb
Capella
Castor
Pollux
Procyon
Betelgeuse
Aldebaran
Sirius
Regulus
Arcturus
Spica
Antares
Rigil Kentaurus
Hadar
Acrux
Gacrux
Mimosa
Avior
Regor
Suhail
Alphard
Denebola
Algieba
Dubhe
Merak
Phecda
Megrez
Alioth
Mizar
Alcor
Alkaid
Thuban
Kochab
Cor Caroli
Alphecca
Izar
Vindemiatrix
Porrima
Gienah
Alchiba
Menkent
Muhlifain
Zubeneschamali
Zubenelgenubi
Dschubba
Sabik
Rasalhague
Rasalgethi
Alderamin
Albireo
Sadr
Schedar
Caph
Ruchbah
Mirfak
Algol
Almach
Menkalinan
Elnath
Alhena
Wezen
Adhara
Acubens
Eltanin

JUNE

THE MAN WITH THE SNAKE

ON this month's all-sky chart, follow the celestial equator east from the meridian (the dashed line running between due north and due south). Beyond Virgo you will find the large, "compound" constellation of Ophiuchus (Oh-fee-YOU-kus), the Serpent Holder. In *The New Patterns in the Sky* — a superb multicultural book about constellation lore — the late Julius Staal writes that this odd name derives from the Greek roots *ophis* (serpent) and *cheiro-o* (to handle).

I call the Serpent Holder pattern compound because, though originally one, the man and his snake are now two separate constellations: Ophiuchus and Serpens. Furthermore, Serpens is composed of two parts: Serpens (Caput) and Serpens (Cauda), meaning, respectively, head and tail; the snake's midsection is blocked from view by Ophiuchus's body. Thus, Serpens is the sky's only discontinuous constellation.

The words *Caput* and *Cauda* are put in parentheses because without them these two names would be ungrammatical in Latin — and all constellations are treated as Latin names. Guy Ottewell, creator of the renowned *Astronomical Calendar* and Latin savant, underscores that the correct word forms without the parentheses would be *Serpentis Caput* and *Serpentis Cauda*.

Ophiuchus has a fuzzy early mythology and is often linked to the ancient Mesopotamian Sun god Marduk and the dragon Tiamat (a supposed predecessor to the serpent). In Greek lore he is often portrayed as the physician Aesculapius — variants of which include Asclepius, Esculapius, and others — with his snake. Serpents have long been regarded as sources of healing, a concept that has regularly been a leading tenet of medical quackery via "snake-oil elixirs" and such. On the other hand, Aesculapius's serpent has been incorporated into the caduceus — that universal medical emblem with one or two snakes coiled around a rod.

A super physician, Aesculapius mastered the ultimate medical skill: reviving the dead. This ability alarmed Pluto, god of the underworld, who felt he would soon be deprived of customers. So he appealed to the head god, Zeus, to intercede. Wanting his subordinates to continue performing their specialized duties, Zeus assented to Pluto's appeal and struck down Aesculapius. Yet, to recognize his special abilities, Zeus gave him a permanent place in the sky. Another, though less spectacular, feat of this superhealer was reassembling Hippolytus, who had been tied to wild horses and dismembered. I wonder if these tales of Aesculapius's mysterious powers inspired Mary Shelley to create Dr. Frankenstein!

INTERESTING OBJECTS

Object	Constellation	Type	Right Ascension	Dec.	Size/Sep./ Period	Magnitude
Coma Cluster	Coma Ber.	OC	12^h 25^m	+26° 00′	4°.5	1.8
M3	Canes Ven.	GC	13^h $42^m.2$	+28° 23′	16′	6.4
M5	Serpens	GC	15^h $18^m.6$	+02° 05′	17′	5.8
M51 (Whirlpool Galaxy)	Canes Ven.	G	13^h $29^m.9$	+47° 12′	11′ × 8′	8.1
M53	Coma Ber.	GC	13^h $12^m.9$	+18° 10′	13′	7.7
M64 (Black-eye Galaxy)	Coma Ber.	G	12^h $56^m.7$	+21° 41′	9′ × 5′	8.5
M81	Ursa Major	G	09^h $55^m.6$	+69° 04′	26′ × 14′	6.8
M82	Ursa Major	G	09^h $55^m.8$	+69° 41′	11′ × 5′	8.4
M94	Canes Ven.	OC	12^h $50^m.9$	+41° 07′	11′ × 9′	8.1
M104 (Sombrero Gal.)	Virgo	G	12^h $40^m.0$	–11° 37′	9′ × 4′	8.3
M106	Canes Ven.	G	12^h $19^m.0$	+47° 18′	18′ × 8′	8.3
NGC 6543	Draco	PN	17^h $58^m.6$	+66° 38′	5′.8	9
ε Boötis (Izar)	Boötes	DS	14^h $45^m.0$	+27° 04′	2″.8	2.5, 4.9
R Coronae Borealis	Corona Bor.	VS	15^h $48^m.6$	+28° 09′	irregular	5.7–14.8
ζ Ursae Majoris (Mizar)	Ursa Major	DS	13^h $23^m.9$	+54° 56′	AB 14″.4 AC 709″	2.3, 4.0 4.0

DN = diffuse nebula; DS = double or multiple star; DkN = dark nebula; G = galaxy; GC = globular cluster; OC = open cluster; PN = planetary nebula; VS = variable star

M5 in Serpens

M104 in Virgo

Notice how Ophiuchus is directly above one of the sky's deadliest creatures, the Scorpion, who, among other things, stung the mighty hunter Orion to death. Indeed, our celestial physician could not have been more appropriately located, given his unusual skills. This reinforces my belief that the placement of constellations in the sky was not completely random but partly due to the relations among neighboring groups (see the essay for March, Northern Hemisphere).

Scorpius was supposedly placed as far as possible from Orion so that the creature could cause him no further trouble. Ophiuchus standing on the Scorpion could thus be called "anti-Orion," because he is located directly opposite the mighty Hunter on the celestial sphere. In other words, Ophiuchus appears in the June evening sky just about where Orion is found a half year later at the same time of night. One distinction between Orion and Ophiuchus, however, is that the former is the name of a specific mythological individual while the latter is not.

By the way, our physician friend also had two daughters: Hygeia and Panacea. It's obvious how their names became familiar health-related words. Interestingly, Ophiuchus and his daughters have long been patron saints of sorts to modern-day doctors. In fact, the Hippocratic oath, named after the very real ancient Greek physician Hippocrates (about 460–377 B.C.) — often regarded as the "father of medicine" — begins, "I swear by Apollo Physician and Aesculapius and Hygeia and Panacea and all the gods and goddesses, making them my witness, that I will fulfill according to my ability and judgment this oath and covenant."

— *George Lovi*

M106 in Canes Venatici

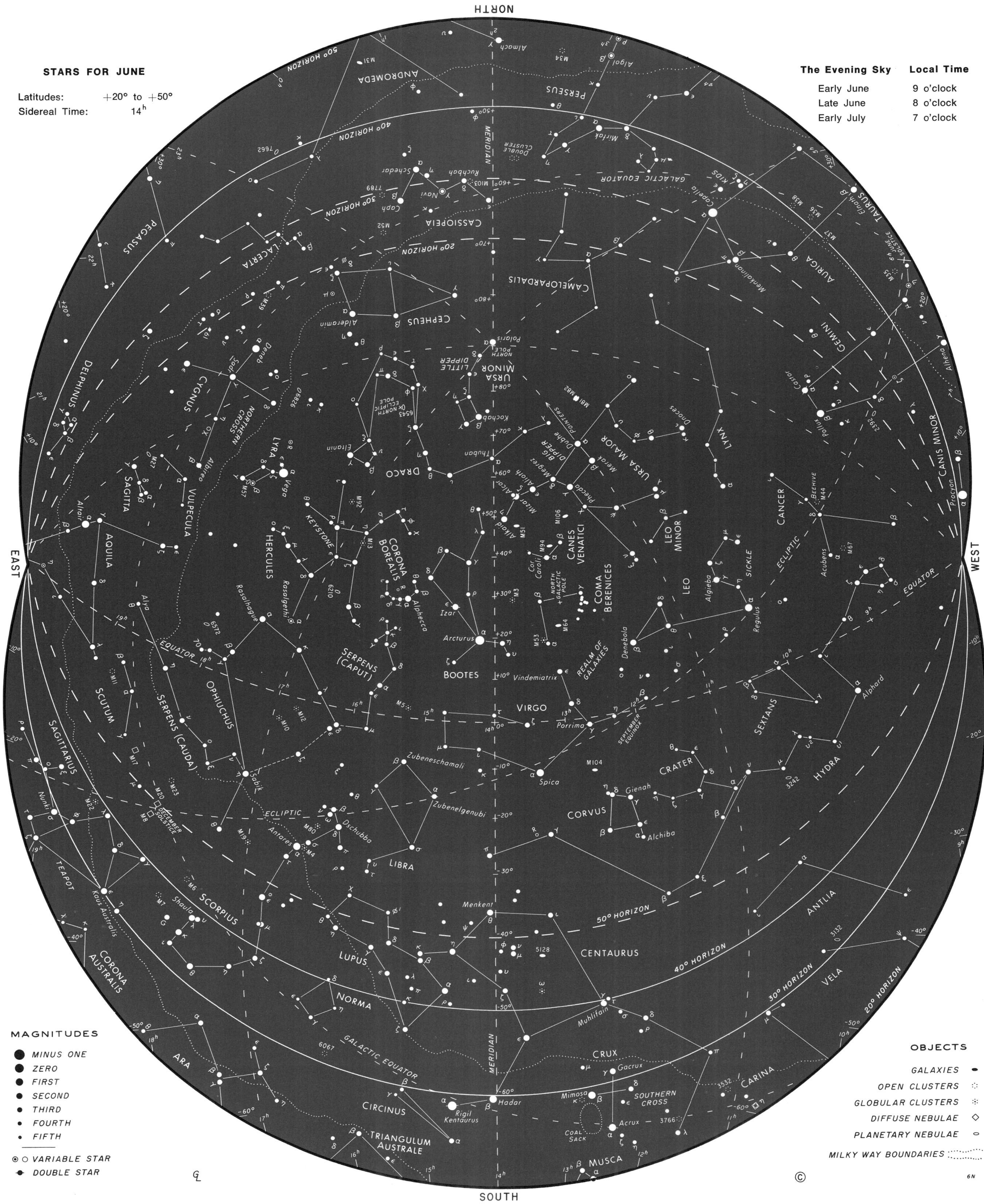
STARS FOR JUNE
Latitudes: +20° to +50°
Sidereal Time: 14h
The Evening Sky Local Time
Early June 9 o'clock
Late June 8 o'clock
Early July 7 o'clock
NORTH
SOUTH
EAST
WEST
MAGNITUDES
MINUS ONE
ZERO
FIRST
SECOND
THIRD
FOURTH
FIFTH
VARIABLE STAR
DOUBLE STAR
OBJECTS
GALAXIES
OPEN CLUSTERS
GLOBULAR CLUSTERS
DIFFUSE NEBULAE
PLANETARY NEBULAE
MILKY WAY BOUNDARIES

JULY

Legends of Herculean Proportions

MANY of our oldest constellations, I've noted before, can be traced to the beginnings of recorded history in the Middle East, specifically Mesopotamia. This is the region around the Tigris and Euphrates Rivers that has become today's Iraq.

One of the best-known star groups that developed from Mesopotamian culture at least five millennia ago is high overhead this month after dusk. The ancient hero we call by his Latin name, Hercules, has long represented a man of extraordinary strength. He appears in various forms in the legends of many peoples throughout the region, including the tale of Samson in the Bible.

Perhaps the earliest identifiable Hercules predecessor is the Sumerian hero Gilgamesh. Like Hercules, he was partially a god, being only half mortal. But Gilgamesh was also the mythological ruler of the Sumerian city Erech (the site of the modern Warka, Iraq) and a rather harsh, tyrannical one. Not unlike the way Hercules was assigned 12 superhuman labors, Gilgamesh had to perform the awesome task of single-handedly reconstructing and restoring the buildings and temples of the city.

Gilgamesh's lust for power greatly displeased the god Anu. To subdue our Sumerian hero, Anu sent a wild half-human, half-beast creature called Enkidu to clash with him. After a violent battle Gilgamesh proved victorious. Enkidu's character then changed, and he became Gilgamesh's admirer and faithful companion. The two set off on a series of adventures somewhat like those of Hercules. One tale even involved the slaughter of a savage lion, just as with Hercules and Samson.

These feats enchanted the love goddess, Ishtar, who offered herself to Gilgamesh as his wife. He rejected her, primarily because of her questionable fidelity, since she had a long history of offering herself to others. Incensed that any man could spurn a woman as irresistible as she regarded herself, Ishtar sent a vicious bull (which some have linked to Taurus) to destroy him. When our champion easily dispatched the bull, the goddess afflicted both Gilgamesh and Enkidu with a deadly disease. Gilgamesh, being half god, was immune, but Enkidu, a mortal, succumbed.

Much of the rest of the Gilgamesh epic involves attempts by our now grieving hero to revive his deceased companion. In one set of adventures he visits the underworld, much as later Greek mythological figures do or try to do. During his subterranean sojourn, Gilgamesh seeks out Utnapishtim, the sole survivor of the great flood that inundated the region in an earlier time and who was said to possess the secrets of eternal youth and immortality. Here we have what has long been regarded as the Mesopotamian version of Noah and the flood, since Utnapishtim tells Gilgamesh how the gods in their rage sent a massive deluge to destroy all life but spared him by directing him to construct a boat.

M13 in Hercules

NGC 6543 in Draco

Although Utnapishtim was not granted immortality, he was given the secret of a magic plant that could restore youth. This he revealed to Gilgamesh, who obtained the plant only to have it snatched away by a serpent (note the parallel to the Garden of Eden with its serpent and tree of knowledge) who did not wish anyone else to have the secret of youth and renewal that the serpents had reserved for themselves. This reflects the ancient belief in the healing power of snakes and their associated medical symbolism, which I discussed in last month's essay.

Our own Hercules character is largely a creation of the Greeks, who called him Heracles. The extensive legends surrounding him are among the best known of Greek mythology. We call him Hercules in keeping with the tradition of using Latin names for the constellations. Many Greek gods, heroes, heroines, and other legendary personalities were adopted by the Romans, who identified them with characters of their own. Thus Zeus became Jupiter; Hera, Juno; Ares, Mars; and so forth. When astronomical bodies are given mythological names in modern times (as were the planetary satellites discovered by the Voyager spacecraft), the Latin version has generally been preferred.

The great Roman author, statesman, and philosopher Lucius Annaeus Seneca (4 B.C.–A.D. 65) relates in verse some of the leading Hercules legends in *Hercules Furens* (Mad Hercules) and *Hercules Oetaeus* (Hercules on Oeta), though there is some doubt about Seneca's authorship of the second work.

Hercules Furens includes a diatribe by Juno against Hercules — her husband's son by another woman. She also decries Jupiter's other lovers and resulting offspring and describes how the sky has been "defiled" by some of this crew being placed there. I often think that such tales of intrigue make these ancient legends the soap operas of the sky!

— George Lovi

Interesting Objects

Object	Constellation	Type	Right Ascension	Dec.	Size/Sep./ Period	Magnitude
M4	Scorpius	GC	$16^h\ 23^m.6$	−26° 32′	26′	5.9
M5	Serpens	GC	$15^h\ 18^m.6$	+02° 05′	17′	5.8
M10	Ophiuchus	GC	$16^h\ 57^m.1$	−04° 06′	15′	6.6
M12	Ophiuchus	GC	$16^h\ 47^m.2$	−01° 57′	14′	6.6
M13 (Hercules Cluster)	Hercules	GC	$16^h\ 41^m.7$	+36° 28′	17′	5.9
M51 (Whirlpool Galaxy)	Canes Ven.	G	$13^h\ 29^m.9$	+47° 12′	11′ × 8′	8.1
M80	Scorpius	GC	$16^h\ 17^m.0$	−22° 59′	9′	7.2
M92	Hercules	GC	$17^h\ 17^m.1$	+43° 08′	11′	6.5
NGC 6210	Hercules	PN	$16^h\ 44^m.5$	+23° 49′	0′.2	9
NGC 6543	Draco	PN	$17^h\ 58^m.6$	+66° 38′	5′.8	9
ε Boötis (Izar)	Boötes	DS	$14^h\ 45^m.0$	+27° 04′	2″.8	2.5, 4.9
ν Draconis	Draco	DS	$17^h\ 32^m.2$	+55° 11′	61″.9	4.9, 4.9
α Herculis (Rasalgethi)	Hercules	DS	$17^h\ 14^m.6$	+14° 23′	4″.7	2.5, 5.4
β Scorpii	Scorpius	DS	$16^h\ 05^m.4$	−19° 48′	13″.6	2.6, 4.9
ζ Ursae Majoris (Mizar)	Ursa Major	DS	$13^h\ 23^m.9$	+54° 56′	AB 14″.4 AC 709″	2.3, 4.0 4.0

DN = diffuse nebula; DS = double or multiple star; DkN = dark nebula; G = galaxy; GC = globular cluster; OC = open cluster; PN = planetary nebula; VS = variable star

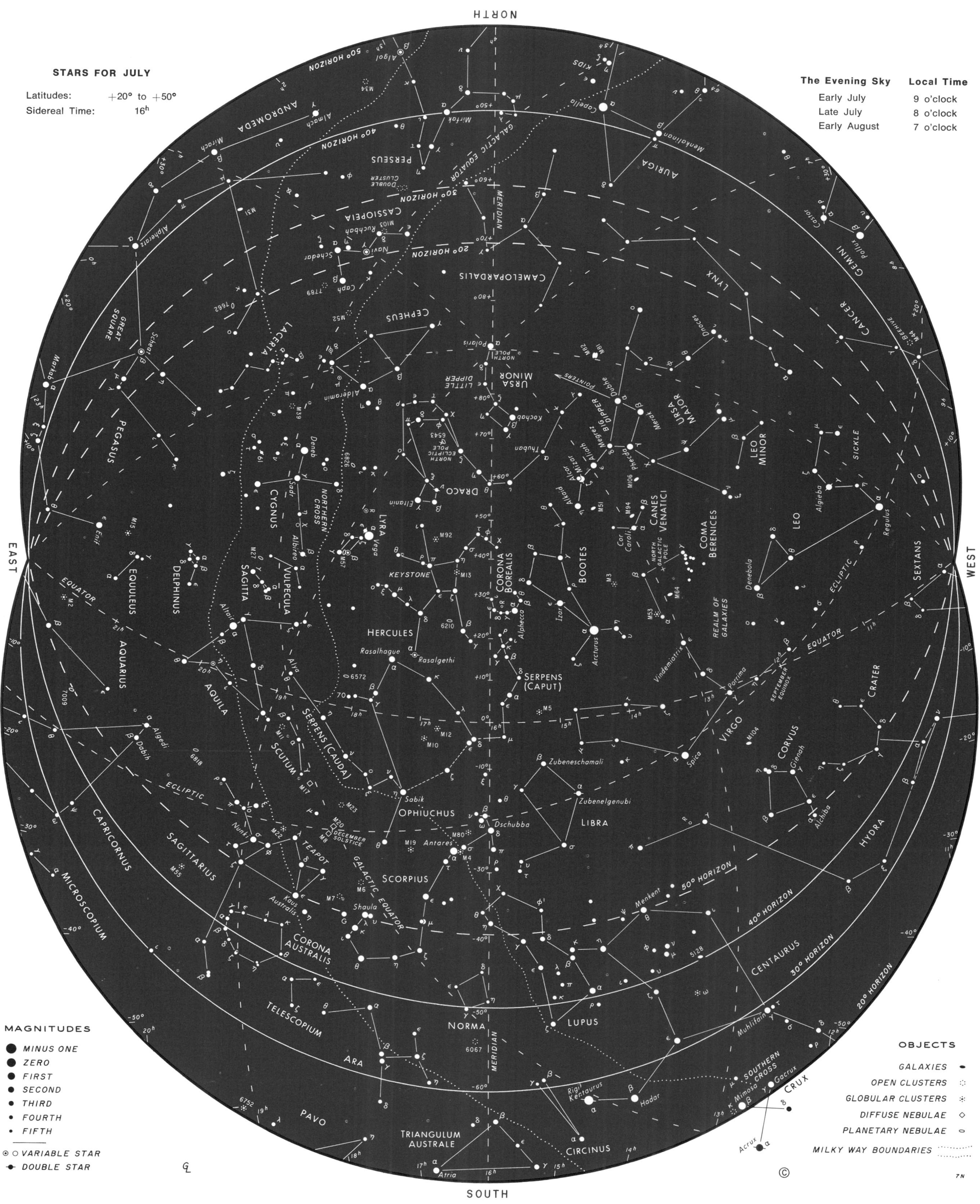

NORTH
SOUTH
EAST
WEST
STARS FOR JULY
Latitudes: +20° to +50°
Sidereal Time: 16h
The Evening Sky
Local Time
Early July
9 o'clock
Late July
8 o'clock
Early August
7 o'clock
MAGNITUDES
MINUS ONE
ZERO
FIRST
SECOND
THIRD
FOURTH
FIFTH
VARIABLE STAR
DOUBLE STAR
OBJECTS
GALAXIES
OPEN CLUSTERS
GLOBULAR CLUSTERS
DIFFUSE NEBULAE
PLANETARY NEBULAE
MILKY WAY BOUNDARIES
ANDROMEDA
PERSEUS
AURIGA
CASSIOPEIA
CAMELOPARDALIS
CEPHEUS
LYNX
GEMINI
CANCER
LACERTA
URSA MINOR
LITTLE DIPPER
URSA MAJOR
BIG DIPPER
POINTERS
LEO MINOR
SICKLE
PEGASUS
GREAT SQUARE
CYGNUS
NORTHERN CROSS
LYRA
DRACO
CANES VENATICI
COMA BERENICES
LEO
SEXTANS
EQUULEUS
DELPHINUS
SAGITTA
VULPECULA
KEYSTONE
CORONA BOREALIS
BOOTES
NORTH GALACTIC POLE
REALM OF GALAXIES
HERCULES
AQUARIUS
AQUILA
SERPENS (CAUDA)
SERPENS (CAPUT)
VIRGO
CRATER
CORVUS
SCUTUM
OPHIUCHUS
LIBRA
HYDRA
CAPRICORNUS
SAGITTARIUS
TEAPOT
SCORPIUS
MICROSCOPIUM
CORONA AUSTRALIS
CENTAURUS
TELESCOPIUM
NORMA
LUPUS
ARA
PAVO
TRIANGULUM AUSTRALE
CIRCINUS
SOUTHERN CROSS
CRUX
EQUATOR
ECLIPTIC
GALACTIC EQUATOR
MERIDIAN
DECEMBER SOLSTICE
SEPTEMBER EQUINOX
NORTH POLE
NORTH ECLIPTIC POLE
50° HORIZON
40° HORIZON
30° HORIZON
20° HORIZON

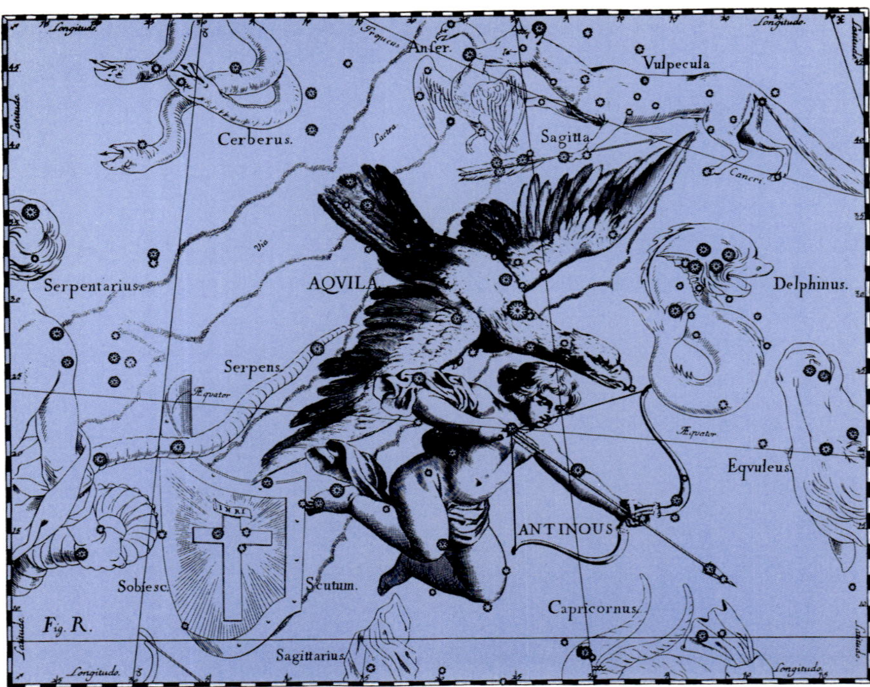

AUGUST

Milky Way Meanderings

IN August the richest part of the Milky Way stretches across the eastern and southern heavens. Regrettably, too many of us can't see it from the cities and suburbs where we live. To see our galactic wreath in its actual splendor you might have to travel far to dark skies, away from the glow of light pollution. In late summer, the farther south you are the better. But be sure to check the phases of the Moon when planning such an outing. Try to avoid the two weeks of each month when it's bright in the evening, since moonlight obscures the Milky Way from even the best viewing locations.

The term *Milky Way* is truly ancient. On old star charts the ghostly band was often labeled by its Latin equivalent, *Via Lactea*. However, if I had my druthers, I'd call it the Smoky Way. To me it looks more smoky than milky, especially with those summer star clouds.

Certain early Mideastern peoples thought so, too. To them, the clouds represented smoke billowing from Ara, the altar, just beneath Scorpius. (Many centuries ago, the precessional shift of the heavens placed these constellations farther north and more easily visible from the Mideast.) To Australia's aborigines the misty appearance of the Milky Way is smoke reflecting the light of stellar campfires. Today, the Milky Way's brightest section, just above the spout of Sagittarius's "Teapot," suggests a puff of vapor emerging from that more modern contrivance.

A leading Greek legend also suggests the origin of the Milky Way. Zeus, king of the gods, placed his half-mortal son Heracles against the breast of his sleeping immortal wife, Hera. The infant, better known by his Roman name Hercules, drank with such strength that the milk spilled across the heavens and created the *Via Lactea.*

As astronomical knowledge grew during the past century, the term *Milky Way* was applied by extension to the galaxy in which we live. So keep in mind the distinction between the Milky Way *band,* which is projected against the sky, and the Milky Way *galaxy,* a three-dimensional structure in space.

The galaxy appears confined to a narrow band because we're looking along its plane. But that explains nothing about the vagaries of its appearance — its often chaotic brightness, width, and texture. These characteristics offer valuable clues to the structure of our galaxy, particularly the area around our Sun.

Let's take an imaginary trip through the galaxy to see how it relates to the hazy swath in the sky. For this we borrow a set of diagrams from *The Beginner's Guide to the Skies* by C. H. Cleminshaw, a superb, fact-filled work that explains many astronomical hows and whys.

M17 in Sagittarius

In the diagram on this page, we venture off through interstellar space and look back at ever-larger spheres or "bubbles" centered on the Sun. Their radii are 10, 100, 1,000, and 5,000 light-years at A, B, C, and D, respectively. The number of stars assigned to each sphere is merely suggestive, illustrating how star counts grow exponentially with larger volumes of space. These diagrams assume that the stars are evenly distributed — which is definitely not the case but an approximation to get the basic idea across.

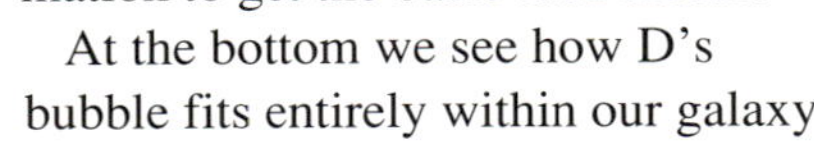

At the bottom we see how D's bubble fits entirely within our galaxy viewed edge on. The galaxy's disk is actually much less than 10,000 light-years thick by most definitions. Various populations of stars and other objects define different disk thicknesses. For instance, nebulae and extremely young stars tend to lie within just a few hundred light-years of the Milky Way's plane.

Almost all naked-eye stars — ones forming our constellations — fall within diagram C (1,000 light-years). The few we can see beyond this bubble include the brilliant blue supergiants that make up the magnificent constellations from Orion eastward to Scorpius. Since they generally follow the Milky Way band, these stars show a preference for our galactic plane. The nearer naked-eye stars, on the other hand, those within sphere B, are more uniformly spread across the heavens.

When we use a telescope and seek out stars beyond the 5,000-light-year radius of sphere D, we find that they predominantly congregate in the galactic plane. Such stars, individually too faint to see with the unaided eye, are the ones that form our Smoky Way. But since there's an awful lot of obscuring interstellar matter and debris out there as well, even a telescope can show only a limited number of distant stars. If this interstellar veil were suddenly lifted, the Sagittarius star clouds would be so bright that even city dwellers could see them!

— *George Lovi*

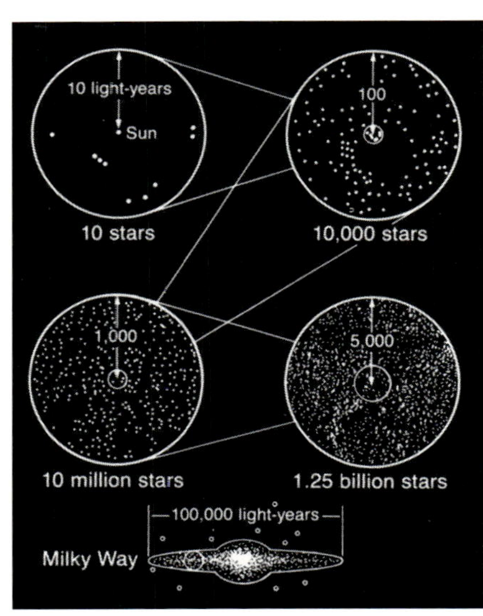

Centered on the Sun, these circles represent successively greater volumes of space. As distance increases, the stars blend to form the band of the Milky Way.

Interesting Objects

Object	Constellation	Type	Right Ascension	Dec.	Size/Sep./ Period	Magnitude
M6 (Butterfly Cluster)	Scorpius	OC	$17^h\ 40^m$	−32° 13′	15′	4.2
M7	Scorpius	OC	$17^h\ 53^m.9$	−34° 49′	80′	3.3
M8 (Lagoon Nebula)	Sagittarius	DN	$18^h\ 03^m.8$	−24° 23′	90′ × 40′	6
M10	Ophiuchus	GC	$16^h\ 57^m.1$	−04° 06′	15′	6.6
M11 (Wild Duck Cl.)	Scutum	OC	$18^h\ 51^m.1$	−06° 16′	14′	5.8
M12	Ophiuchus	GC	$16^h\ 47^m.2$	−01° 57′	14′	6.6
M13 (Hercules Cluster)	Hercules	GC	$16^h\ 41^m.7$	+36° 28′	17′	5.9
M17 (Omega Nebula)	Sagittarius	DN	$18^h\ 20^m.8$	−16° 11′	46′ × 37′	7
M19	Ophiuchus	GC	$17^h\ 02^m.6$	−26° 16′	14′	7.2
M20 (Trifid Nebula)	Sagittarius	DN	$18^h\ 02^m.6$	−23° 02′	29′ × 27′	8
M22	Sagittarius	GC	$18^h\ 36^m.4$	−23° 54′	24′	5.1
M23	Sagittarius	OC	$17^h\ 56^m.8$	−19° 01′	27′	5.5
M57 (Ring Nebula)	Lyra	PN	$18^h\ 53^m.6$	+33° 02′	1′	9
M92	Hercules	GC	$17^h\ 17^m.1$	+43° 08′	11′	6.5
NGC 6210	Hercules	PN	$16^h\ 44^m.5$	+23° 49′	0′.2	9
NGC 6543	Draco	PN	$17^h\ 58^m.6$	+66° 38′	5′.8	9
NGC 6826 (Blinking Nebula)	Cygnus	PN	$19^h\ 44^m.8$	+50° 31′	2′	10

DN = diffuse nebula; DS = double or multiple star; DkN = dark nebula; G = galaxy; GC = globular cluster; OC = open cluster; PN = planetary nebula; VS = variable star

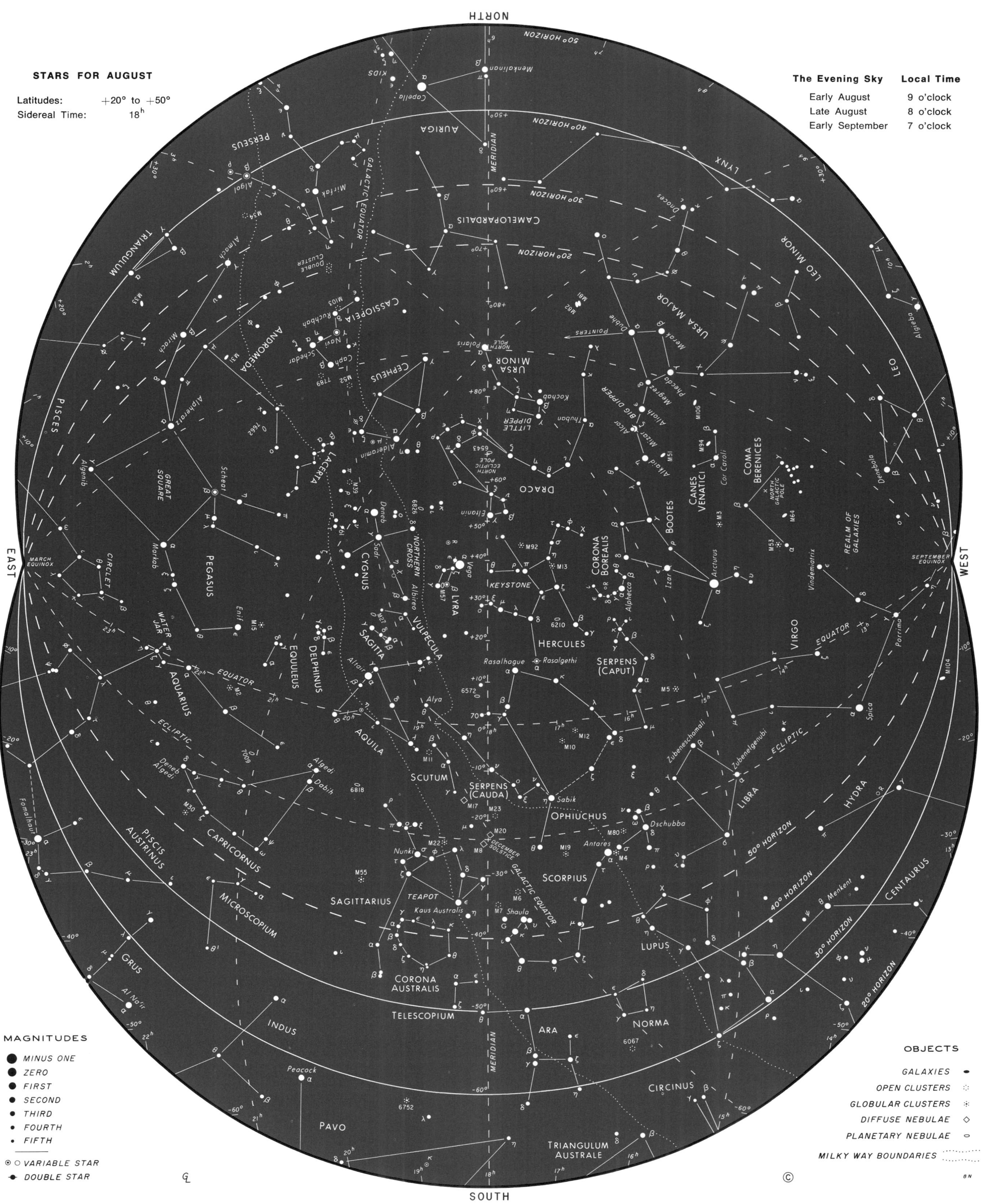
STARS FOR AUGUST
Latitudes: +20° to +50°
Sidereal Time: 18h
The Evening Sky Local Time
Early August 9 o'clock
Late August 8 o'clock
Early September 7 o'clock
NORTH
SOUTH
EAST
WEST
MAGNITUDES
MINUS ONE
ZERO
FIRST
SECOND
THIRD
FOURTH
FIFTH
VARIABLE STAR
DOUBLE STAR
OBJECTS
GALAXIES
OPEN CLUSTERS
GLOBULAR CLUSTERS
DIFFUSE NEBULAE
PLANETARY NEBULAE
MILKY WAY BOUNDARIES
AURIGA
PERSEUS
CAMELOPARDALIS
LYNX
LEO MINOR
URSA MAJOR
TRIANGULUM
CASSIOPEIA
ANDROMEDA
CEPHEUS
URSA MINOR
LITTLE DIPPER
BIG DIPPER
DRACO
LEO
PISCES
GREAT SQUARE
LACERTA
COMA BERENICES
CANES VENATICI
BOOTES
CORONA BOREALIS
HERCULES
KEYSTONE
LYRA
CYGNUS
NORTHERN CROSS
PEGASUS
CIRCLET
VULPECULA
SAGITTA
DELPHINUS
EQUULEUS
WATER JAR
AQUARIUS
SERPENS (CAPUT)
VIRGO
REALM OF GALAXIES
AQUILA
SCUTUM
SERPENS (CAUDA)
OPHIUCHUS
LIBRA
HYDRA
CAPRICORNUS
PISCIS AUSTRINUS
SAGITTARIUS
TEAPOT
SCORPIUS
MICROSCOPIUM
CORONA AUSTRALIS
LUPUS
CENTAURUS
GRUS
TELESCOPIUM
ARA
NORMA
INDUS
PAVO
CIRCINUS
TRIANGULUM AUSTRALE
EQUATOR
ECLIPTIC
GALACTIC EQUATOR
MERIDIAN
MARCH EQUINOX
SEPTEMBER EQUINOX
DECEMBER SOLSTICE
NORTH ECLIPTIC POLE
NORTH GALACTIC POLE
NORTH POLE
POINTERS
50° HORIZON
40° HORIZON
30° HORIZON
20° HORIZON

SEPTEMBER

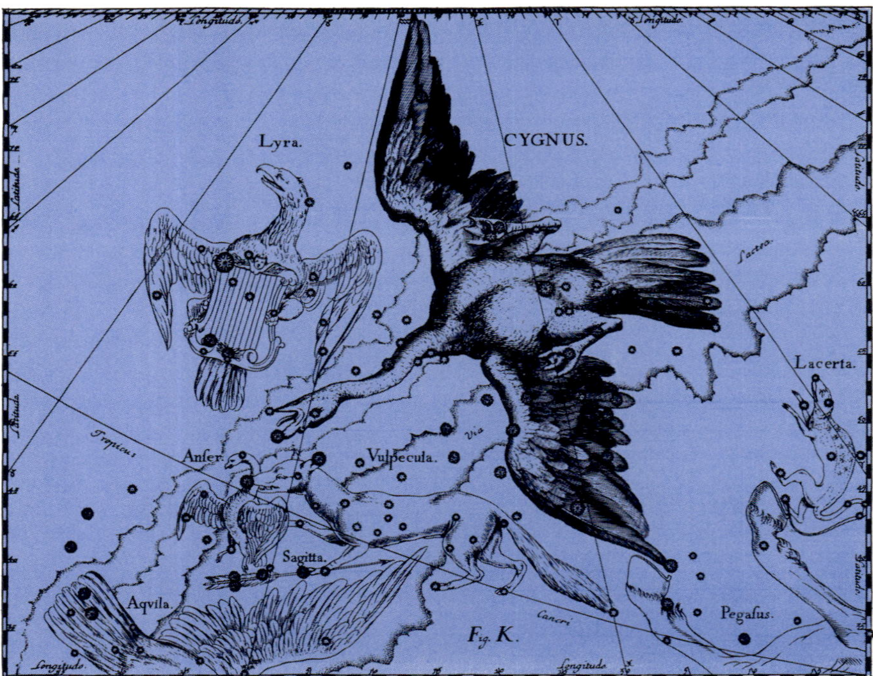

FIGURING HOW FAR TO A STAR

SCATTERED across the firmament are certain dim stars that hold special significance all out of proportion to their brightnesses. One of them can be found just east of the center of this month's chart: 61 Cygni. This 5th-magnitude member of the Swan forms a parallelogram with Deneb, Sadr, and Epsilon (ε) Cygni, the northern and eastern segments of the Northern Cross.

Why is this speck of light special? Because 61 Cygni was the first star whose distance was measured directly, using the exacting method of trigonometric parallax. Vega and Alpha Centauri came in a close second and third.

Friedrich Wilhelm Bessel (1784–1846) was working at Königsberg Observatory in the then easternmost corner of Germany (or Prussia) when he announced the stellar parallax of 61 Cygni in 1838. This momentous event marked one of the true beginnings of stellar astronomy and astrophysics — finding the distances to the stars.

This breakthrough occurred when positional astronomy was already approaching today's standards of precision. Such an achievement 1½ centuries ago was due to exquisite refinements in the art of instrument making, as well as newly rigorous methods of observing and data reduction that Bessel himself helped develop. From this period on astronomers could measure star positions to within a fraction of an arc second, the realm wherein all stellar parallaxes lie.

Astronomers had begun to search for stellar parallax some three centuries before Bessel. They reasoned that if the Earth moved around the Sun as Copernicus claimed, stars should appear to move from side to side during the course of a year. There were many failed attempts and false alarms — spurious parallax measurements — a few of them absurdly large.

Yet all the work was not wasted. Other major discoveries were byproducts of early parallax attempts. In the 1720s British astronomer James Bradley found the aberration of starlight during one of his tries. This phenomenon causes the apparent position of a star, the direction its light seems to be coming from, to be displaced 20.5 arc seconds directly ahead of the Earth's motion. Thus every star during the course of a year describes an elliptical path 41 arc seconds across against the sky. (For a star on the ecliptic this ellipse becomes a straight line; at the ecliptic poles it's a circle.) The aberration of starlight, many times larger than any stellar parallax, is one of the major corrections that must be applied to a star's observed position.

When William Herschel decided to join the parallax hunt at the start of the last century, he used relatively close double stars and watched for any changes between the components. Instead of finding a parallax, he noticed that a few of his doubles were progressively changing their separations and orientations in a way that proved the components were orbiting each other under the influence of gravity. This was a major breakthrough in itself, revealing that the law of gravitation extended beyond the solar system.

M39 in Cygnus

M27 in Vulpecula

Another early 18th-century discovery — proper motion, by Edmond Halley — helped set the stage for the successful parallax measurements a century later. Proper motion is the slow change in a star's position over time, making our familiar constellations transient phenomena. What alerted Bessel to the possibility that 61 Cygni was a good parallax candidate was its unusually large proper motion of 5.2 arc seconds per year. He reasoned that this meant the star is relatively nearby, and so he commenced his painstaking measurements of its position.

Bessel looked for a "weave" in 61 Cygni's proper-motion path having a period of one year. He succeeded in measuring the star's parallax with a now obsolete position-measuring telescope called a heliometer, which used a split objective lens whose halves moved independently and thus could form a double image to measure small angular displacements. Although it was not easy to use, the instrument remained the preferred parallax-measuring tool for the first dozen or so stars throughout the 1800s. Just after 1900, far more precise and effective photographic procedures revolutionized the art and made parallax measurement much more productive both in quality and quantity.

Even as Bessel was making his observations, his German colleague Friedrich Wilhelm Struve was obtaining a successful parallax of Vega with the 9-inch Dorpat refractor just up the Baltic coast. He used a filar eyepiece micrometer — with essentially the same techniques he used for all his high-quality double star observations.

The third parallax obtained, for Alpha (α) Centauri, was secured by the Scottish astronomer Thomas Henderson (1798–1844) using a now-obsolete meridian transit instrument called a mural circle. He, too, was alerted by unusually large proper motion. It was the very large parallax of Alpha Centauri — still the greatest of any star's — that was chiefly responsible for Henderson's success.

— George Lovi

INTERESTING OBJECTS

Object	Constellation	Type	Right Ascension	Dec.	Size/Sep./ Period	Magnitude
M2	Aquarius	GC	$21^h\ 33^m.5$	−00° 49′	13′	6.5
M11 (Wild Duck Cl.)	Scutum	OC	$18^h\ 51^m.1$	−06° 16′	14′	5.8
M15	Pegasus	GC	$21^h\ 30^m.0$	+12° 10′	12′	6.4
M27 (Dumbbell Nebula)	Vulpecula	PN	$19^h\ 59^m.6$	+22° 43′	8′ × 4′	8
M39	Cygnus	OC	$21^h\ 32^m.2$	+48° 26′	32′	4.6
M57 (Ring Nebula)	Lyra	PN	$18^h\ 53^m.6$	+33° 02′	1′	9
NGC 6543	Draco	PN	$17^h\ 58^m.6$	+66° 38′	5′.8	9
NGC 6818	Sagittarius	PN	$19^h\ 44^m.0$	−14° 09′	0′.3	10
NGC 6826 (Blinking Nebula)	Cygnus	PN	$19^h\ 44^m.8$	+50° 31′	2′	10
NGC 7009 (Saturn Nebula)	Aquarius	PN	$21^h\ 04^m.2$	−11° 22′	2′	8
α Capricorni (Algedi)	Capricorn	DS	$20^h\ 18^m.1$	−12° 33′	378″	3.6, 4.2
β Cygni (Albireo)	Cygnus	DS	$19^h\ 30^m.7$	−27° 58′	34″.4	3.1, 5.1
61 Cygni	Cygnus	DS	$21^h\ 06^m.9$	+38° 45′	29″.7	5.2, 6.0
β Lyrae (Sheliak)	Lyra	VS	$18^h\ 50^m.1$	+33° 22′	12.9 days	3.34–4.34
R Lyrae	Lyra	VS	$18^h\ 55^m.3$	+43° 57′	46.0 days	3.88–5.0

DN = diffuse nebula; DS = double or multiple star; DkN = dark nebula; G = galaxy; GC = globular cluster; OC = open cluster; PN = planetary nebula; VS = variable star

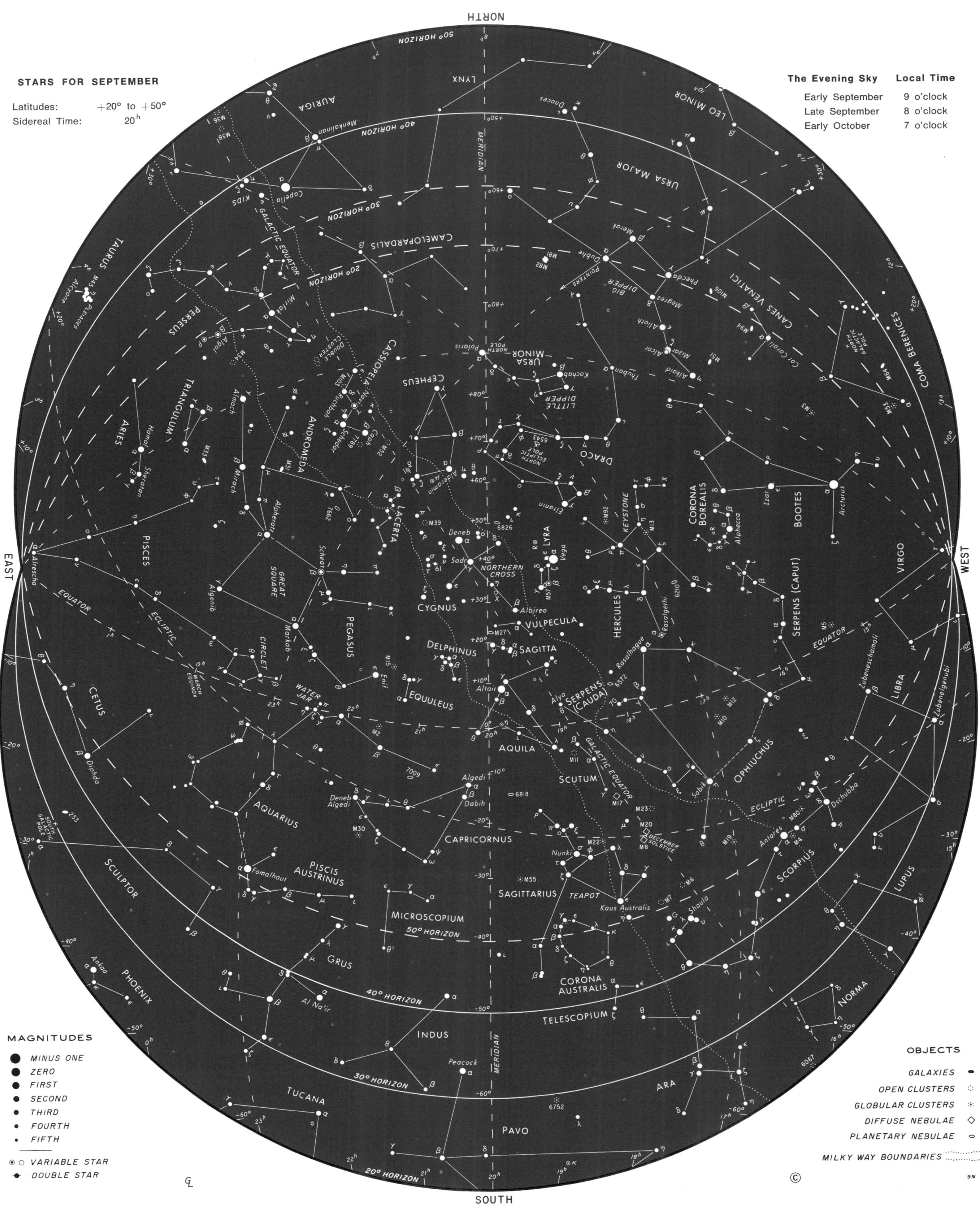
STARS FOR SEPTEMBER
Latitudes: +20° to +50°
Sidereal Time: 20h
The Evening Sky
Local Time
Early September 9 o'clock
Late September 8 o'clock
Early October 7 o'clock
NORTH
SOUTH
EAST
WEST
LYNX
AURIGA
LEO MINOR
URSA MAJOR
CAMELOPARDALIS
TAURUS
PERSEUS
CANES VENATICI
COMA BERENICES
CASSIOPEIA
URSA MINOR
CEPHEUS
LITTLE DIPPER
BIG DIPPER
TRIANGULUM
ARIES
ANDROMEDA
DRACO
LACERTA
CORONA BOREALIS
BOOTES
KEYSTONE
PISCES
GREAT SQUARE
LYRA
NORTHERN CROSS
CYGNUS
HERCULES
SERPENS (CAPUT)
VIRGO
PEGASUS
VULPECULA
DELPHINUS
SAGITTA
CIRCLET
EQUULEUS
SERPENS (CAUDA)
LIBRA
CETUS
WATER JAR
AQUILA
OPHIUCHUS
SCUTUM
AQUARIUS
CAPRICORNUS
SCORPIUS
PISCIS AUSTRINUS
SCULPTOR
SAGITTARIUS
TEAPOT
LUPUS
MICROSCOPIUM
GRUS
CORONA AUSTRALIS
PHOENIX
NORMA
TELESCOPIUM
INDUS
ARA
TUCANA
PAVO
MERIDIAN
EQUATOR
ECLIPTIC
GALACTIC EQUATOR
50° HORIZON
40° HORIZON
30° HORIZON
20° HORIZON
MAGNITUDES
MINUS ONE
ZERO
FIRST
SECOND
THIRD
FOURTH
FIFTH
VARIABLE STAR
DOUBLE STAR
OBJECTS
GALAXIES
OPEN CLUSTERS
GLOBULAR CLUSTERS
DIFFUSE NEBULAE
PLANETARY NEBULAE
MILKY WAY BOUNDARIES
©

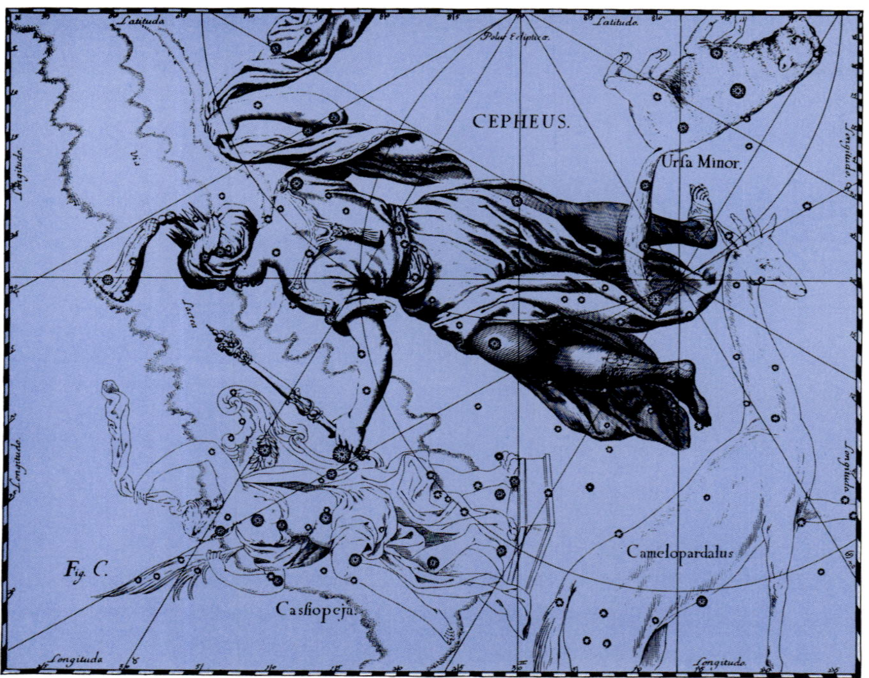

OCTOBER

Following Those Variable Stars

ASTRONOMY is almost unique as a science in that even in this high-tech era the ordinary lay person, or amateur, can still make significant contributions. Variable star observing is one example of a worthwhile activity for the backyard enthusiast.

The sky has many thousands of stars whose brightnesses vary by amounts from a small fraction of a magnitude to 10 magnitudes or more, with periods anywhere from a few hours to a few years. Variable stars come in a multitude of categories. A major distinction is that some are true, or intrinsic, variables, while others are extrinsic ones. The latter include eclipsing binaries, in which the component stars pass in front of each other, blocking one another's light and thus lessening the total light we see on a regular cycle.

An intrinsic variable, on the other hand, alters its light output because of cyclic physical changes in the star itself, such as the pulsations that may begin late in the star's life span. One major category here is the Cepheid variables, which vary in brightness as much as two magnitudes, with periods ranging from less than two days to more than two months. Another comprises long-period variables, which fluctuate by several magnitudes over three months to a year or more. The long-period performers, usually red giants, are watched by a large coterie of amateurs. Their light curves never repeat exactly in amplitude or period. Cepheids are far more regular on both counts.

This month's sky map contains three well-known prototype variables. About halfway up in the northeast is the famous eclipsing binary Algol in Perseus, whose 1.3-magnitude dimming every 69 hours can be readily followed with the unaided eye. (*Sky & Telescope*'s Calendar Notes section carries a schedule of Algol's minima each month.) Just east of the meridian high in the north is the best-known Cepheid, Delta (δ) Cephei, which varies by less than a magnitude every 5⅓ days. This star, and many more like it, helped astronomers determine the distances to galaxies. And finally, low in the east-southeast we find Mira in Cetus.

Each of these stars is represented by the special variable star symbol that's become virtually universal on star charts and atlases: a concentric circle and dot that roughly approximate the star's magnitude range, or an open circle if the minimum is below the chart's magnitude limit. On the charts in this book only well-known or significant variables are identified in this way. In fact, Polaris is a small-range Cepheid shown as a regular star.

With nothing more than a pair of binoculars, the light changes of hundreds of stars can be followed. These include Mira, the brightest long-period variable. The finder chart on this page shows the field around this star, which has long been used as a practice object for budding variable star observers. Mira typically ranges between magnitude 3.4 and 9.5 over a period of 332 days. However, we cannot always tell what it will do; back in 1779 it brightened to 1st magnitude, the virtual equal of Aldebaran!

M103 in Cassiopeia

Observers estimate Mira's brightness by comparing it to nearby stars. On the finder chart, you'll notice various stars labeled with a two-digit number: 54, 57, 68, and so on. These are the visual magnitudes of convenient comparison stars with the decimal point omitted (to avoid possible confusion with star dots). For example, if Mira appears halfway in brightness between the ones labeled 67 and 71, then the observer records Mira as being magnitude 6.9.

Every now and then an amateur astronomer gets lucky. Back in 1955 I started doing some variable star observing and decided to join the American Association of Variable Star Observers (AAVSO). I included a report of Mira with my application. I made the sighting on a night when the star was an exact match to that 9.2-magnitude star to its immediate east (left); they looked like a beautiful "cat's eyes" double. In response I received a note from an AAVSO staffer, Helen Stephansky, complimenting me on how skilled I was!

The AAVSO is still the leading organization devoted to observing and recording variable stars. Those interested in getting involved or joining can write them at 25 Birch St., Cambridge, MA 02138, U.S.A.

Predawn observations of long-period variables when they are low in the eastern sky are especially valuable, since many have periods close to a year. This causes a perpetual lapse of data in one part of the light curve, owing to solar glare and the fact that most people prefer to observe during the evening. The early-morning observations fill in those important gaps.

Low-range variables such as Cepheids and certain eclipsing binaries are targets for the more advanced amateur with photo-electric equipment. Here a very precise light curve is needed, and the eye just can't judge bright-nesses precisely enough.

But on long-period Mira-type stars a well-trained human eye remains very effective. Among other things, it lets you judge magnitudes directly, without any need for extensive calculations to reduce photoelectric readings into usable data.

— *George Lovi*

γ Ceti 36
North
α Ceti 27
δ Ceti 41
61
54
55
East
60
57
64
73
92 Mira
86
80
58
68
71
67
75
70
3°

Interesting Objects

Object	Constellation	Type	Right Ascension	Dec.	Size/Sep./ Period	Magnitude
M2	Aquarius	GC	21ʰ 33ᵐ.5	−00° 49′	13′	6.5
M15	Pegasus	GC	21ʰ 30ᵐ.0	+12° 10′	12′	6.4
M30	Capricorn	GC	21ʰ 40ᵐ.4	−23° 11′	11′	7.5
M31 (Andromeda Gal.)	Andromeda	G	00ʰ 42ᵐ.7	+41° 16′	178′ × 63′	3.4
M39	Cygnus	OC	21ʰ 32ᵐ.2	+48° 26′	32′	4.6
M52	Cassiopeia	OC	23ʰ 24ᵐ.2	+61° 35′	13′	6.9
M103	Cassiopeia	OC	01ʰ 33ᵐ.2	+60° 42′	6′	7
NGC 6543	Draco	PN	17ʰ 58ᵐ.6	+66° 38′	5′.8	9
NGC 7009 (Saturn Nebula)	Aquarius	PN	21ʰ 04ᵐ.2	−11° 22′	2′	8
NGC 7789	Cassiopeia	OC	23ʰ 57ᵐ.0	+56° 44′	16′	6.7
γ Cassiopeiae (Navi)	Cassiopeia	VS	00ʰ 56ᵐ.7	+60° 43′	irregular	1.6–3.0
δ Cephei	Cepheus	VS	22ʰ 29ᵐ.2	+58° 25′	5.4 days	3.5–4.4
β Persei (Algol)	Perseus	VS	03ʰ 08ᵐ.2	+40° 57′	2.9 days	2.1–3.4

DN = diffuse nebula; DS = double or multiple star; DkN = dark nebula; G = galaxy; GC = globular cluster; OC = open cluster; PN = planetary nebula; VS = variable star

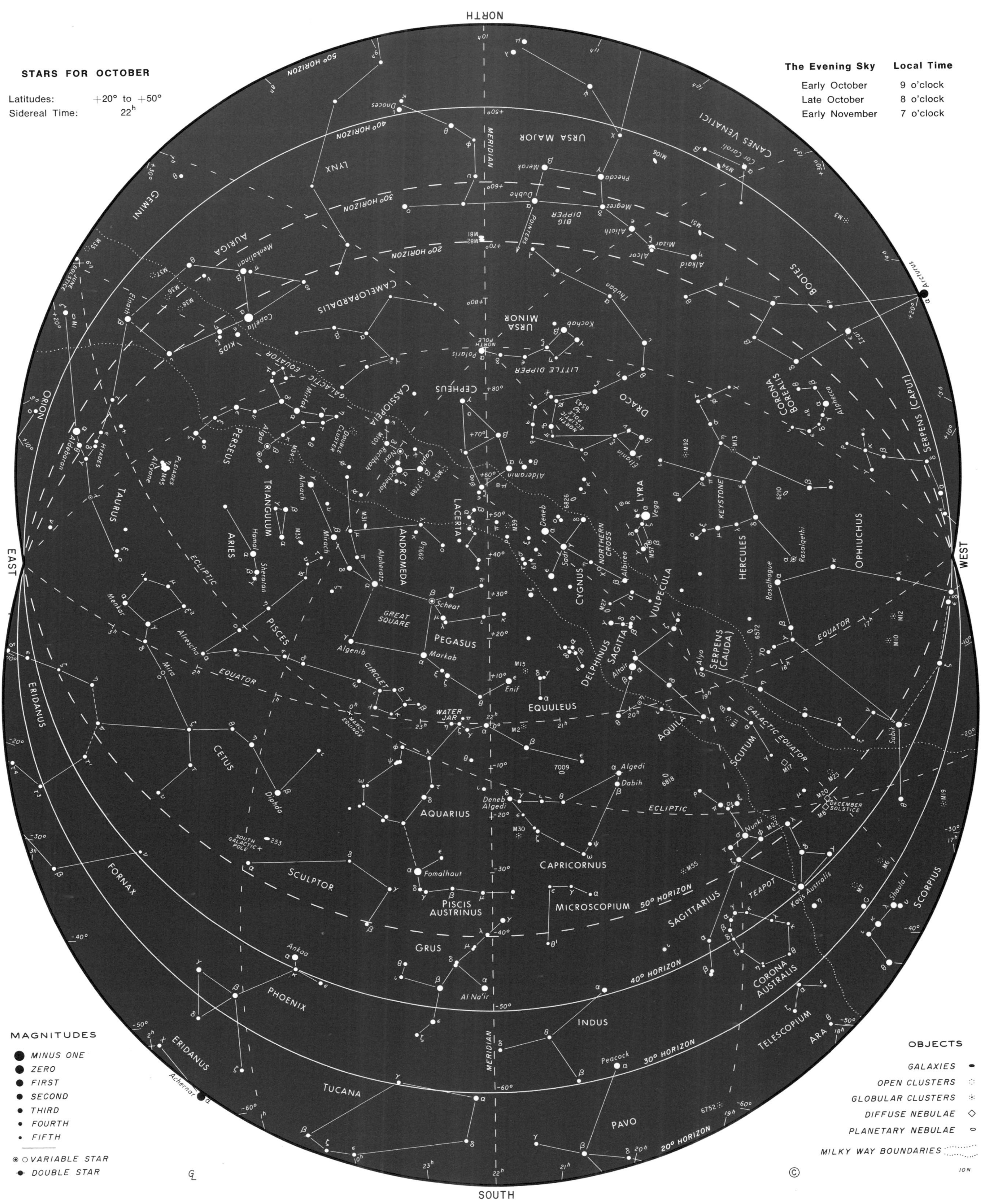
STARS FOR OCTOBER
Latitudes: +20° to +50°
Sidereal Time: 22h
The Evening Sky Local Time
Early October 9 o'clock
Late October 8 o'clock
Early November 7 o'clock
NORTH
SOUTH
EAST
WEST
URSA MAJOR
LYNX
CANES VENATICI
GEMINI
AURIGA
CAMELOPARDALIS
BOOTES
URSA MINOR
LITTLE DIPPER
BIG DIPPER
CEPHEUS
DRACO
CORONA BOREALIS
ORION
PERSEUS
CASSIOPEIA
SERPENS (CAPUT)
TAURUS
TRIANGULUM
ARIES
ANDROMEDA
LACERTA
LYRA
HERCULES
OPHIUCHUS
CYGNUS
NORTHERN CROSS
VULPECULA
SAGITTA
DELPHINUS
PEGASUS
GREAT SQUARE
PISCES
SERPENS (CAUDA)
EQUULEUS
AQUILA
SCUTUM
ERIDANUS
CETUS
AQUARIUS
CAPRICORNUS
SCULPTOR
FORNAX
PISCIS AUSTRINUS
MICROSCOPIUM
SAGITTARIUS
TEAPOT
SCORPIUS
GRUS
PHOENIX
CORONA AUSTRALIS
INDUS
TELESCOPIUM
ARA
TUCANA
PAVO
ECLIPTIC
EQUATOR
GALACTIC EQUATOR
MERIDIAN
MAGNITUDES
MINUS ONE
ZERO
FIRST
SECOND
THIRD
FOURTH
FIFTH
VARIABLE STAR
DOUBLE STAR
OBJECTS
GALAXIES
OPEN CLUSTERS
GLOBULAR CLUSTERS
DIFFUSE NEBULAE
PLANETARY NEBULAE
MILKY WAY BOUNDARIES

NOVEMBER

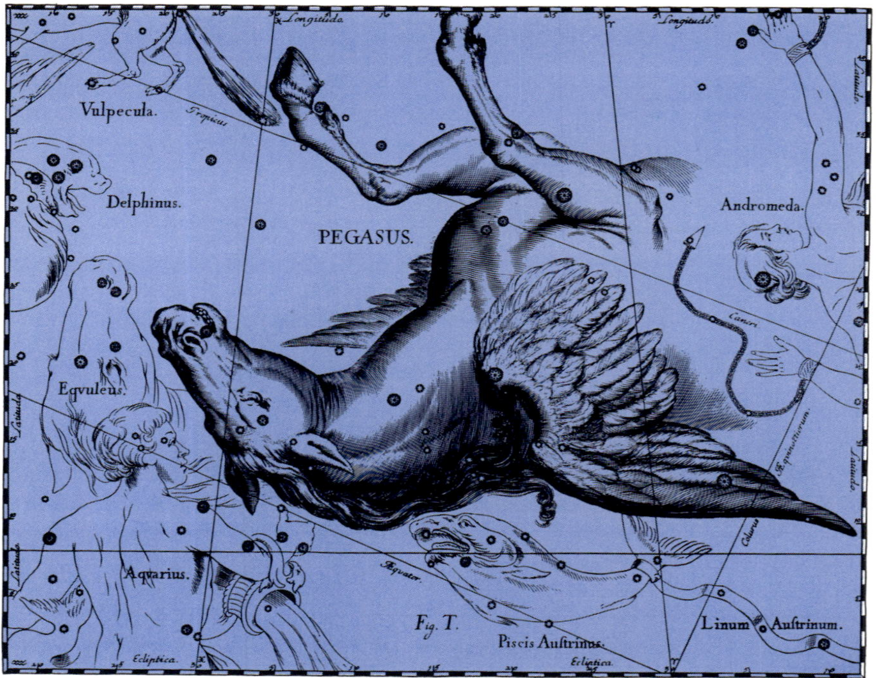

PEREGRINATIONS WITH PEGASUS

A FAMILIAR ancient constellation occupies a commanding position high in the south on this month's chart: the flying horse, Pegasus, leader of the fall constellations parade. It is one of the easier patterns for the beginner to trace, since its stars form a fairly good figure of the front half of a galloping steed. His nose is marked by the star Enif, and his front legs stretch out from Scheat at the upper right (northwest) corner of the prominent Great Square asterism that forms the horse's body.

However, the animal is upside down in the sky to most Northern Hemisphere observers, as is the legendary strong man Hercules, now disappearing in the northwest sky. While visiting New Zealand one June, I had the opportunity to see Pegasus right side up just before dawn in Rotorua. In addition to the constellation's "correct" orientation, its apparent enlargement due to the "Moon illusion" effect (since it was near the horizon) made it a most impressive sight.

This is just one part of what makes stargazing from drastically different latitudes so much fun. It's not just seeing stars unobservable from back home. Even familiar constellations seem to take on new aspects when presented differently, with risings and settings, paths across the sky, and orientations quite unlike what we're used to. Just as a familiar musical composition sounds fresh and different when played in a new arrangement, so do constellations look fresh and different when newly arranged. Although, as I mentioned in my essay for the month of July, Hercules appeared right side up to the early Mideast peoples more than five millennia ago, such was not the case with Pegasus. The creature was then located right on the celestial equator; if anything, it appeared upside down from even more places on Earth.

Like several other constellations, such as Capricornus, Centaurus, Draco, Hydra, Monoceros, and Phoenix, Pegasus represents a mythical beast. The most common account of his origin lies in the legend of how Perseus beheaded the Gorgon Medusa; the Winged Horse sprang from her blood. This makes him part of the supporting cast of the Royal Family of the Sky drama that includes King Cepheus, Queen Cassiopeia, Princess Andromeda, hero and suitor Perseus, and sea monster Cetus. They are all very favorably placed in the November evening sky.

According to some versions of the story, Pegasus's first task was to carry Perseus to Ethiopia. There Andromeda was chained to a seacoast rock as Cetus approached to devour her. Perseus arrived just in time. He held up Medusa's head, snakes and all, for Cetus to gaze at, which turned the creature into stone while Perseus carried off his bride-to-be.

NGC 869/884, the Perseus Double Cluster

Pegasus was subsequently involved in further adventures. He normally resided on Mount Olympus, abode of the gods, and came to Earth infrequently. On one such occasion the youth Bellerophon was assigned the task of slaughtering another of those ancient mythical composite monsters, the Chimera, which had a lion's head, goat's body, and dragon's tail. This is the origin of our word *chimera,* a fantastic combination of incongruous parts or the result of wild imagining.

Bellerophon spent an entire night in the temple of Minerva, imploring help from the goddess in carrying out the task. She decided to lend him her aid; the next morning Bellerophon woke to find a golden bridle in his hand and Pegasus drinking from a nearby fountain.

The young lad approached the majestic animal, which allowed itself to be harnessed and mounted. With the aid of this steed, Bellerophon dispatched the Chimera with relative ease. But, as mythological characters are wont to do, he became too proud for his own good. Pegasus seemed to give him godlike powers, and the youth began to regard himself as more than an ordinary mortal. After several heady adventures with Pegasus, Bellerophon decided to visit the Flying Horse's home base, Mount Olympus, to meet some of its other inhabitants. The gods were astounded at the young man's audacity as horse and rider approached the Olympian gates. Jupiter sent a gadfly to sting Pegasus, who bolted and threw his rider. When Bellerophon fell to Earth he was rendered blind and lame, a condition in which he wandered about for the rest of his days.

M33 in Triangulum

A nonmythological footnote to the tale concerns Napoleon Bonaparte — a real-life individual who perhaps thought he, too, was more than mortal. The ship that carried Napoleon to his final exile on the island of St. Helena was named the *Bellerophon.*

— George Lovi

INTERESTING OBJECTS

Object	Constellation	Type	Right Ascension	Dec.	Size/Sep./ Period	Magnitude
M31 (Andromeda Galaxy)	Andromeda	G	$00^h\ 42^m.7$	+41° 16′	178′ × 63′	3.4
M33 (Pinwheel Galaxy)	Triangulum	G	$01^h\ 33^m.9$	+30° 39′	62′ × 39′	5.7
M39	Cygnus	OC	$21^h\ 32^m.2$	+48° 26′	32′	4.6
M52	Cassiopeia	OC	$23^h\ 24^m.2$	+61° 35′	13′	6.9
M103	Cassiopeia	OC	$01^h\ 33^m.2$	+60° 42′	6′	7
NGC 253 (Sculptor Galaxy)	Sculptor	G	$12^h\ 43^m.7$	+11° 33′	7′ × 6′	8.8
NGC 869/884	Perseus	OC	$02^h\ 19^m.0$	+57° 09′	30′	4
(Double Cluster)		OC	$02^h\ 22^m.4$	+57° 07′	30′	4
NGC 7662 (Blue Snowball)	Andromeda	PN	$23^h\ 25^m.9$	+42° 33′	2′	9
NGC 7789	Cassiopeia	OC	$23^h\ 57^m.0$	+56° 44′	16′	6.7
γ Andromedae	Andromeda	DS	$02^h\ 03^m.9$	+42° 20′	AB 9″.8	2.2, 5.1
(Almach)					BC 0″.5	5.5, 6.3
γ Arietis	Aries	DS	$01^h\ 53^m.5$	+19° 18′	7″.8	4.8, 4.8
γ Cassiopeiae (Navi)	Cassiopeia	VS	$00^h\ 56^m.7$	+60° 43′	irregular	1.6–3.0
δ Cephei	Cepheus	VS	$22^h\ 29^m.2$	+58° 25′	5.4 days	3.5–4.4
o Ceti (Mira)	Cetus	VS	$02^h\ 19^m.3$	–02° 59′	332 days	2.0–10.1
β Persei (Algol)	Perseus	VS	$03^h\ 08^m.2$	+40° 57′	2.9 days	2.12–3.40

DN = diffuse nebula; DS = double or multiple star; DkN = dark nebula; G = galaxy; GC = globular cluster; OC = open cluster; PN = planetary nebula; VS = variable star

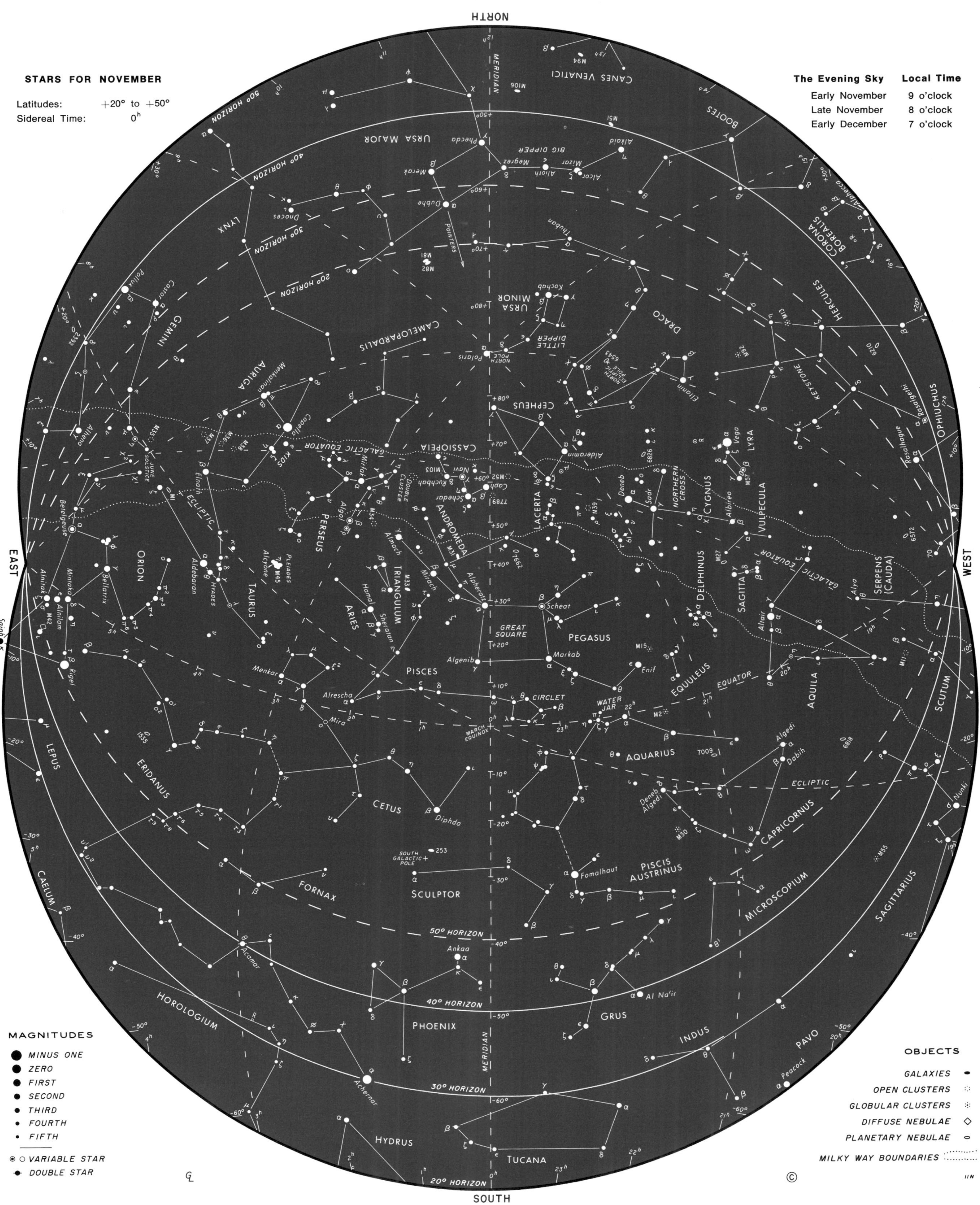
STARS FOR NOVEMBER
Latitudes: +20° to +50°
Sidereal Time: 0h
The Evening Sky
Local Time
Early November 9 o'clock
Late November 8 o'clock
Early December 7 o'clock
NORTH
SOUTH
EAST
WEST
MERIDIAN
MAGNITUDES
MINUS ONE
ZERO
FIRST
SECOND
THIRD
FOURTH
FIFTH
VARIABLE STAR
DOUBLE STAR
OBJECTS
GALAXIES
OPEN CLUSTERS
GLOBULAR CLUSTERS
DIFFUSE NEBULAE
PLANETARY NEBULAE
MILKY WAY BOUNDARIES
CANES VENATICI
URSA MAJOR
BIG DIPPER
BOOTES
CORONA BOREALIS
LYNX
URSA MINOR
LITTLE DIPPER
DRACO
HERCULES
KEYSTONE
CAMELOPARDALIS
GEMINI
AURIGA
CEPHEUS
CASSIOPEIA
LYRA
OPHIUCHUS
NORTHERN CROSS
CYGNUS
VULPECULA
PERSEUS
ANDROMEDA
LACERTA
ORION
TAURUS
TRIANGULUM
ARIES
DELPHINUS
SAGITTA
SERPENS (CAUDA)
GREAT SQUARE
PEGASUS
PISCES
EQUULEUS
AQUILA
SCUTUM
CIRCLET
WATER JAR
AQUARIUS
LEPUS
ERIDANUS
CETUS
CAPRICORNUS
SOUTH GALACTIC POLE
FORNAX
SCULPTOR
PISCIS AUSTRINUS
MICROSCOPIUM
SAGITTARIUS
CAELUM
PHOENIX
GRUS
HOROLOGIUM
INDUS
PAVO
HYDRUS
TUCANA
ECLIPTIC
EQUATOR
GALACTIC EQUATOR
MARCH EQUINOX
JUNE SOLSTICE
NORTH POLE
50° HORIZON
40° HORIZON
30° HORIZON
20° HORIZON

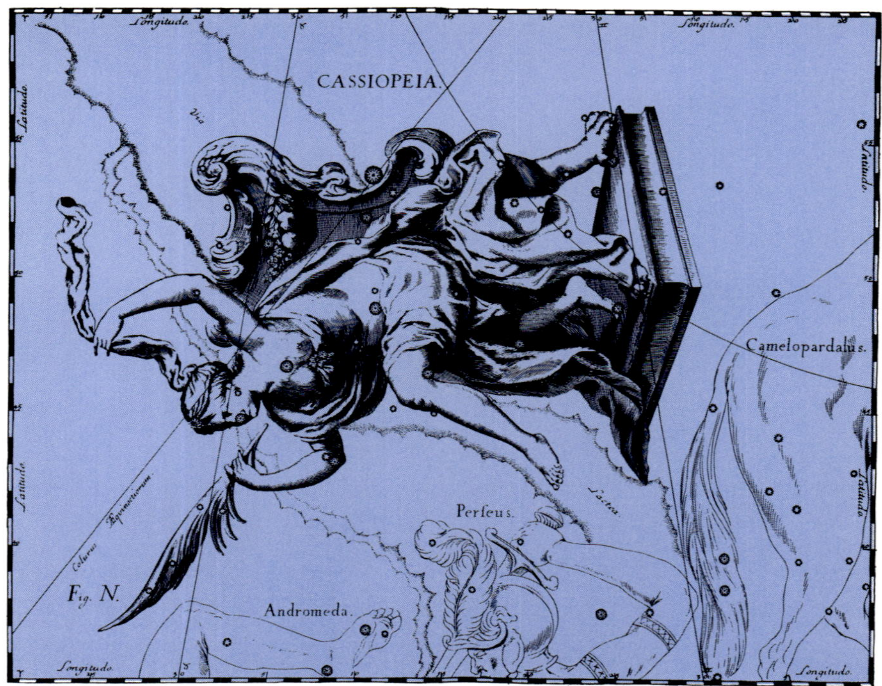

DECEMBER

A Look at Seeing

DECEMBER brings the first crystal clear, sparkling nights of winter. Newcomers to astronomy often assume that winter's especially dark skies mean all forms of observing are at their best. But this is far from so.

The "crystal clear" winter skies and unusually bright stars are due to two factors. First, the cold temperatures prevent the air from holding as much moisture as in summer; less haze and greater atmospheric transparency are often the result. This is also why daytime skies in winter are generally a deeper, crisper blue. Second, the stars overhead at this time of year really are brighter than those of summer. They are among the most luminous beacons lining our local spiral arm of the Milky Way galaxy.

The atmospheric transparency makes winter excellent for binocular viewing, for hunting down faint, elusive nebulae and clusters, and for observing the minima of certain variable stars. These are basically low-power pursuits.

The other side of the coin is that cold, sparkling skies are often very unstable, with strong atmospheric turbulence that makes star images churn and fuzz when viewed at high power. This turbulence is also the reason for the strong stellar scintillation during winter nights; the charming "twinkle, twinkle, little star" of the nursery rhyme is the most obvious evidence of an unstable atmosphere.

Such unsteadiness is called "seeing," and it limits the performance of all ground-based telescopes no matter how large or optically excellent. Rarely can any instrument employ a power greater than 250× to 400× without just magnifying a blur. The problem often looks *worse* in a large telescope.

Interesting Objects

Object	Constellation	Type	Right Ascension	Dec.	Size/Sep./ Period	Magnitude
Hyades	Taurus	OC	$04^h\ 27^m$	+16° 00′	6°	0.5
M31 (Andromeda Galaxy)	Andromeda	G	$00^h\ 42^m.7$	+41° 16′	178′ × 63′	3.4
M33 (Pinwheel Galaxy)	Triangulum	G	$01^h\ 33^m.9$	+30° 39′	62′ × 39′	5.7
M34	Perseus	OC	$02^h\ 02^m.0$	+42° 47′	35′	5.2
M45 (Pleiades)	Taurus	OC	$03^h\ 47^m.0$	+24° 07′	110′	1.2
M52	Cassiopeia	OC	$23^h\ 24^m.2$	+61° 35′	13′	6.9
M103	Cassiopeia	OC	$01^h\ 33^m.2$	+60° 42′	6′	7
NGC 253 (Sculptor Galaxy)	Sculptor	G	$12^h\ 43^m.7$	+11° 33′	7′ × 6′	8.8
NGC 869/884	Perseus	OC	$02^h\ 19^m.0$	+57° 09′	30′	4
(Double Cluster)		OC	$02^h\ 22^m.4$	+57° 07′	30′	4
NGC 7789	Cassiopeia	OC	$23^h\ 57^m.0$	+56° 44′	16′	6.7
γ Arietis	Aries	DS	$01^h\ 53^m.5$	+19° 18′	7″.8	4.8, 4.8
o Ceti (Mira)	Cetus	VS	$02^h\ 19^m.3$	−02° 59′	332 days	2.0–10.1
β Persei (Algol)	Perseus	VS	$03^h\ 08^m.2$	+40° 57′	2.9 days	2.12–3.40
ρ Persei	Perseus	VS	$03^h\ 05^m.2$	+38° 50′	50 days	3.3–4.0

DN = diffuse nebula; DS = double or multiple star; DkN = dark nebula; G = galaxy; GC = globular cluster; OC = open cluster; PN = planetary nebula; VS = variable star

Poor seeing arises where warm and cool air masses meet. Cool air is slightly denser and bends light a little more. Where two temperatures meet, the boundary layer breaks up into a turbulent swirl of warm and cool air pockets typically a few inches across. These disturbances act like lenses and divert light passing through. Since this happens even when the temperature difference is very small, a few hundredths of a degree, the atmosphere is perpetually unsteady. More extreme cases of seeing, visible without optical aid, are the heat waves over a sunlit road or above a fire.

Regular telescopic observers distinguish between "slow" and "fast" seeing. The slow type leaves an image relatively sharp but makes it wobble about. Fast seeing makes the image churn and boil more quickly than the eye can follow.

An important capability of the human eye that cameras and film lack is the ability to pick out detail in a jiggling image. Another is the ability to catch the fleeting moments when seeing improves and the image sharpens up. If you try to photograph your object at such a moment, by the time you press the shutter release it may all be over.

Fast seeing originates high in the sky, and not much can be done about it. But slow seeing is often due to very local, preventable causes. Try to set up your telescope away from sources of warmth, such as houses or pavement that has been warmed by sunlight all day. Grass and trees make excellent surroundings for a telescope; water is even better.

M34 in Perseus

NGC 7789 in Cassiopeia

A related problem that might be mistaken for poor seeing can occur inside the telescope itself. "Tube currents," eddies of warm and cool air, are especially troublesome in large-aperture instruments. Open-tube reflectors suffer the worst when brought outdoors, but they air out and settle down more quickly than closed-tube telescopes, in which minor tube currents can persist for much longer. Look at a bright star out of focus. If the disk is crossed by thin, ropy, bright and dark lines that writhe very slowly, there's a source of warmth in or near the telescope. The solution is to wait until the instrument cools down to the temperature of the surrounding air. Better yet, keep it stored in an unheated area. Also, don't let your own breath or body heat get in the light path during wintertime.

Seeing tends to improve later in the evening. It is always worst near the horizon, where we are viewing through more air. Surprisingly, some of the best seeing occurs on warm, hazy nights that look pretty mediocre to the naked eye. When the atmosphere is loaded with haze droplets or smog, its temperature structure tends to make the seeing *quite* a bit better. This is why summer skies are usually better for high-power pursuits than winter nights. Since conditions for best seeing and best transparency tend to be opposite, there's always something to observe regardless of sky conditions.

The late Lawrence Braymer, founder of Questar Corp., once commented that the best seeing he ever knew occurred near Pittsburgh back when industrial smoke was poorly controlled and the Moon was yellow even when high in the sky. Indeed, Pittsburgh has a major observatory within its city limits, Allegheny. This venerable institution has long been a world leader in precise positional work, such as measuring parallax and proper motion of stars — a field in which seeing is more important than a dark sky.

— George Lovi

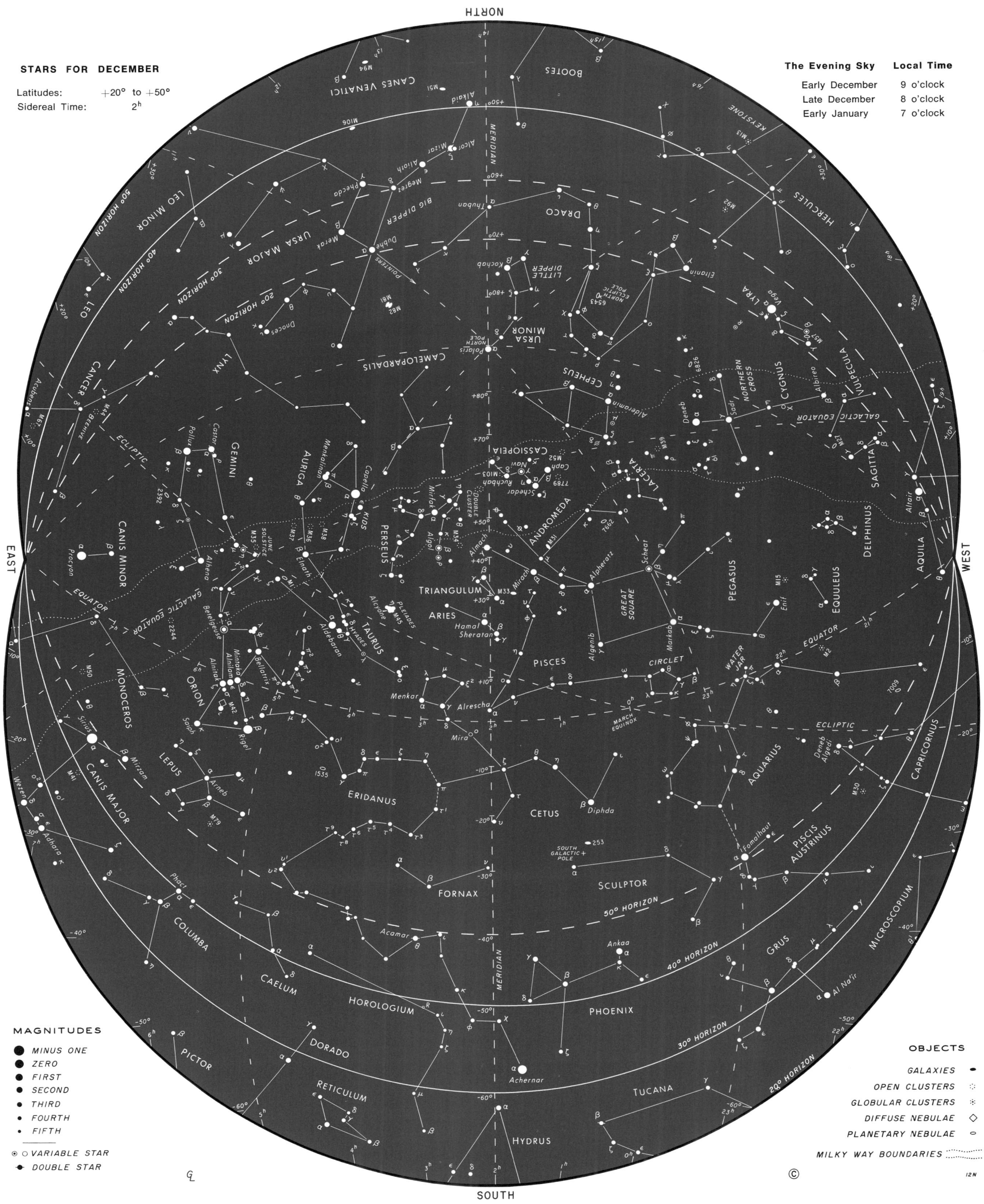
STARS FOR DECEMBER
Latitudes: +20° to +50°
Sidereal Time: 2h
The Evening Sky | Local Time
Early December 9 o'clock
Late December 8 o'clock
Early January 7 o'clock
NORTH
SOUTH
EAST
WEST
MAGNITUDES
MINUS ONE
ZERO
FIRST
SECOND
THIRD
FOURTH
FIFTH
VARIABLE STAR
DOUBLE STAR
OBJECTS
GALAXIES
OPEN CLUSTERS
GLOBULAR CLUSTERS
DIFFUSE NEBULAE
PLANETARY NEBULAE
MILKY WAY BOUNDARIES
CANES VENATICI
BOOTES
KEYSTONE
HERCULES
LEO MINOR
URSA MAJOR
BIG DIPPER
DRACO
LITTLE DIPPER
URSA MINOR
LYRA
LEO
CANCER
LYNX
CAMELOPARDALIS
CEPHEUS
CYGNUS
NORTHERN CROSS
VULPECULA
GEMINI
AURIGA
CASSIOPEIA
LACERTA
SAGITTA
CANIS MINOR
PERSEUS
ANDROMEDA
DELPHINUS
AQUILA
PEGASUS
EQUULEUS
TRIANGULUM
ARIES
TAURUS
GREAT SQUARE
ORION
MONOCEROS
PISCES
CIRCLET
WATER JAR
ECLIPTIC
EQUATOR
GALACTIC EQUATOR
MERIDIAN
LEPUS
CANIS MAJOR
ERIDANUS
CETUS
AQUARIUS
CAPRICORNUS
PISCIS AUSTRINUS
FORNAX
SCULPTOR
COLUMBA
CAELUM
HOROLOGIUM
PHOENIX
GRUS
MICROSCOPIUM
PICTOR
DORADO
RETICULUM
TUCANA
HYDRUS
50° HORIZON
40° HORIZON
30° HORIZON
20° HORIZON
SOUTH GALACTIC POLE
NORTH POLE
NORTH ECLIPTIC POLE
MARCH EQUINOX
JUNE SOLSTICE
Polaris
Capella
Aldebaran
Betelgeuse
Rigel
Sirius
Procyon
Pollux
Castor
Vega
Deneb
Altair
Achernar
Fomalhaut
Mira
Algol
Hamal
Diphda
Ankaa
Acamar

SOUTHERN HEMISPHERE

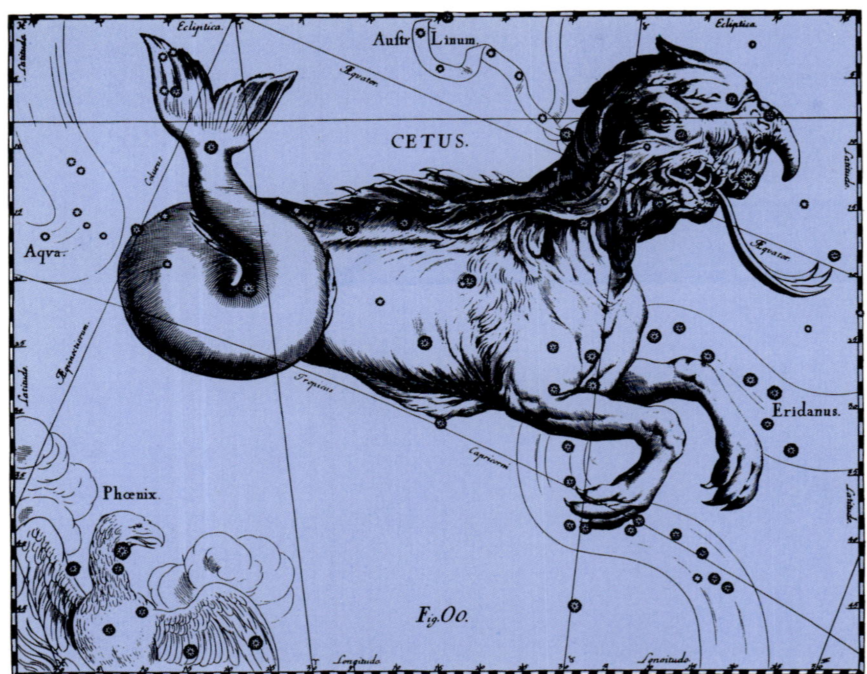

January

Magellanic "Toolhouses of Great Merit"

THE sudden appearance of a supernova in the Large Magellanic Cloud in February of 1987 turned the world's attention to this small, special galaxy in our Milky Way's backyard. I heard groans to the effect: *Why* did it have to happen so near the south celestial pole, out of sight from most of the Northern Hemisphere? And *why* couldn't it have been just a year earlier, when hordes of astronomy buffs had traveled south to see Halley's Comet?

The placement of the exploding star reinforced a certain paranoia that long-time astronomy enthusiasts sometimes develop. In a corollary to Murphy's Law, the heavens seem to conspire to make the best things unobservable for the greatest number of people. Consider the "geography" of the heavens. The skies north of the celestial equator don't hold a candle to what lies below. For starters, there's that splendid span of brilliant stars stretching from Orion through Vela, Carina, and Centaurus to Scorpius. They form part of Gould's Belt, named after the 19th-century American astronomer Benjamin Apthorp Gould. Other dazzling sights include the brilliant Crux-Centaurus area; the Milky Way's hub in Sagittarius; the globular cluster Omega Centauri; the best dark nebula, the Coalsack; and one of the two best bright nebulae, Eta Carinae.

Two more unique visual treats of southern declinations are the "satellite" galaxies of our Milky Way. The late Harvard astronomer Harlow Shapley referred to them as "astronomical toolhouses of great merit" — the Large and Small Magellanic Clouds (LMC and SMC).

These companion galaxies are named after the 16th-century Portuguese navigator who commanded the first expedition around the globe (though he himself was killed in the Philippines). Yet Magellan was hardly the first European to notice these clouds, which resemble pieces of the Milky Way that have drifted loose. Decades earlier, other Portuguese navigators regularly sailed south of the equator, particularly around the Cape of Good Hope; they called these smudges the "Cape Clouds." In former times, the LMC and SMC were often referred to by their Latin names, Nubecula Major and Nubecula Minor — Big Cloudlet and Little Cloudlet.

The LMC is about 160,000 light-years away, with the SMC another 20,000 light-years farther. The nearest major galaxy, M31 in Andromeda, is more than a dozen times more distant.

As serious astronomical research became established in the Southern Hemisphere, the value of our galactic companions became ever more important. The Clouds of Magellan are separate galaxies close enough to the Earth that a great number of their constituents — stars, clusters, and nebulosity — can be examined individually. It is also fortunate that the two "Nubeculae" are well clear of the plane of our home galaxy with its intervening bright star clouds and dark dust; otherwise we would not see them.

Large Magellanic Cloud

The Clouds are of special interest to astronomers because their composition is similar to regions in the spiral arms of our own Milky Way and other galaxies. More than four decades ago, Walter Baade labeled such areas Population I. They are characterized by brilliant young stars, often occurring in clusters, and the nebulosity from which stars are born. On the other hand, his Population II stars are the older members that congregate in the gas-free regions near our galaxy's core, in its outer halo, and in globular clusters. By analogy, we can think of Population I regions as communities of young people raising families, and areas of Population II as settlements of senior citizens.

Since the LMC is mostly Population I, it's full of young, massive stellar giants that end their brief lives as violent supernovae. Because we have a much clearer line of sight to the Magellanic Clouds than to most of our own galaxy, it's also not so surprising that such an event was first observed out there — even though the LMC has much less than a tenth the stars of the Milky Way. There could have been a dozen supernovae in our galaxy since 1604, the year of the last known one, all hidden behind the detritus in our galaxy's plane.

Small Magellanic Cloud

The LMC has so much merit because it contains a stellar smorgasbord of objects, all of which are essentially at the same distance, making it easy for astronomers to compare their true sizes and luminosities. Just prior to World War I, Henrietta Leavitt of Harvard Observatory began to analyze the light curves of Cepheid variables in the Magellanic Clouds. From her research, she made the momentous discovery that their periods tell their luminosities. By merely observing the period of a Cepheid in a remote galaxy, we can determine its absolute magnitude. By comparing this value to its apparent brightness, we get the star's distance, and consequently that of the galaxy it is in.

Few people have appreciated the numerous Magellanic treasures more than the late Milky Way specialist Bart Bok. When he was at Australia's Mount Stromlo Observatory, woe to the observer or student at a telescope not pointed at these objects when they were favorably placed on a clear moonless night!

— George Lovi

Interesting Objects

Object	Constellation	Type	Right Ascension	Dec.	Size/Sep./ Period	Magnitude
Hyades	Taurus	OC	04^{h} 27^{m}	+16° 00′	6°	0.5
Large Magellanic Cloud	Dorado	G	05^{h} $23^{m}.6$	–69° 45′	11° × 9°	0.6
Small Magellanic Cloud	Tucana	G	00^{h} $52^{m}.7$	–72° 50′	5° × 3°	2.8
47 Tucanae (NGC 104)	Tucana	GC	00^{h} $24^{m}.1$	–72° 05′	30′.9	4
M1 (Crab Nebula)	Taurus	DN	05^{h} $34^{m}.5$	+22° 01′	6′ × 4′	8
M42 (Orion Nebula)	Orion	DN	05^{h} $35^{m}.4$	–05° 27′	66′ × 60′	4
M45 (Pleiades)	Taurus	OC	03^{h} $47^{m}.0$	+24° 07′	110′	1.2
M79	Lepus	GC	05^{h} $46^{m}.7$	–24° 33′	9′	8.0
NGC 1535	Eridanus	PN	04^{h} $14^{m}.2$	–12° 44′	0′.7	10
ο Ceti (Mira)	Cetus	VS	02^{h} $19^{m}.3$	–02° 59′	332 days	2.0–10.1
β Doradus	Dorado	VS	05^{h} $33^{m}.6$	–62° 29′	9.8 days	3.5–4.1
30 Doradus (Tarantula Nebula, NGC 2070)	Dorado	DN	05^{h} $38^{m}.7$	–69° 06′	40′ × 25′	5
θ Eridani (Acamar)	Eridanus	DS	02^{h} $58^{m}.3$	–40° 18′	8″.2	3.4, 4.5
β Persei (Algol)	Perseus	VS	03^{h} $08^{m}.2$	+40° 57′	2.9 days	2.1–3.4

DN = diffuse nebula; DS = double or multiple star; DkN = dark nebula; G = galaxy; GC = globular cluster; OC = open cluster; PN = planetary nebula; VS = variable star

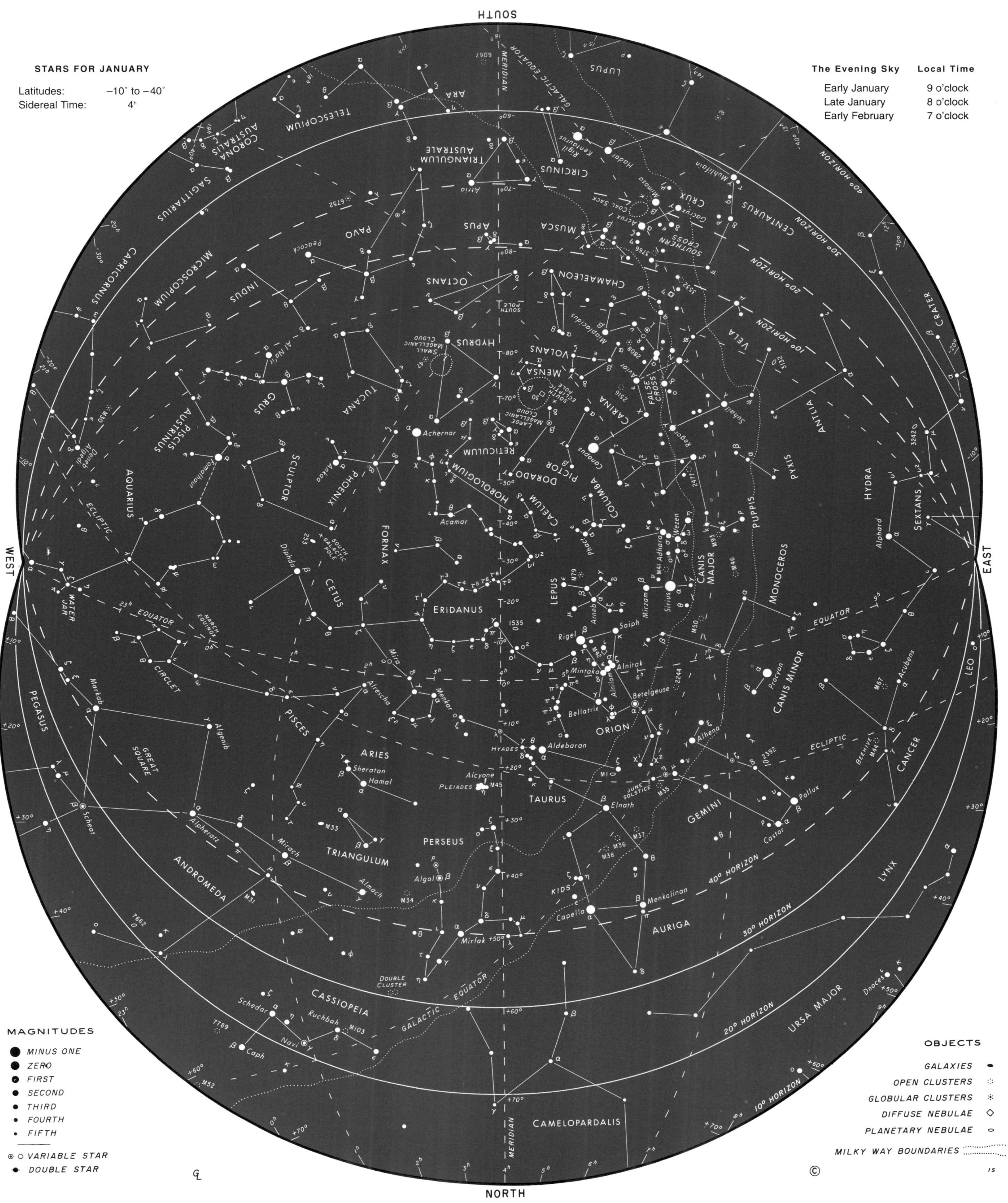
STARS FOR JANUARY
Latitudes: −10° to −40°
Sidereal Time: 4h
The Evening Sky | Local Time
Early January | 9 o'clock
Late January | 8 o'clock
Early February | 7 o'clock
SOUTH
NORTH
EAST
WEST
MERIDIAN
EQUATOR
ECLIPTIC
GALACTIC EQUATOR
10° HORIZON
20° HORIZON
30° HORIZON
40° HORIZON
SOUTH POLE
SOUTH GALACTIC POLE
MARCH EQUINOX
JUNE SOLSTICE
LARGE MAGELLANIC CLOUD
SMALL MAGELLANIC CLOUD
ORION
TAURUS
GEMINI
AURIGA
PERSEUS
CASSIOPEIA
ANDROMEDA
PEGASUS
TRIANGULUM
ARIES
PISCES
CETUS
ERIDANUS
LEPUS
CANIS MAJOR
CANIS MINOR
MONOCEROS
CANCER
LEO
HYDRA
LYNX
URSA MAJOR
CAMELOPARDALIS
PUPPIS
PYXIS
ANTLIA
VELA
CARINA
CENTAURUS
CRUX
MUSCA
CIRCINUS
LUPUS
ARA
TRIANGULUM AUSTRALE
APUS
OCTANS
CHAMAELEON
VOLANS
MENSA
HYDRUS
RETICULUM
DORADO
PICTOR
COLUMBA
CAELUM
HOROLOGIUM
FORNAX
SCULPTOR
PHOENIX
TUCANA
GRUS
INDUS
PAVO
TELESCOPIUM
CORONA AUSTRALIS
SAGITTARIUS
CAPRICORNUS
MICROSCOPIUM
PISCIS AUSTRINUS
AQUARIUS
SEXTANS
CRATER
Sirius
Canopus
Rigel
Betelgeuse
Aldebaran
Capella
Procyon
Pollux
Castor
Achernar
Fomalhaut
MAGNITUDES
MINUS ONE
ZERO
FIRST
SECOND
THIRD
FOURTH
FIFTH
VARIABLE STAR
DOUBLE STAR
OBJECTS
GALAXIES
OPEN CLUSTERS
GLOBULAR CLUSTERS
DIFFUSE NEBULAE
PLANETARY NEBULAE
MILKY WAY BOUNDARIES

FEBRUARY

GREAT STAR OF THE SOUTH

ALTHOUGH February for Northern Hemisphere dwellers means cold winds and winter woollies, for those south of the equator the situation is the opposite. At southern latitudes Orion and his hunting dogs are midsummer constellations, to be appreciated after an evening barbecue or while out walking in the mild summer twilight.

In February, Southern Hemisphere skywatchers have a fine view of the brightest stars in the sky. Sirius is high overhead at midevening as seen from midsouthern latitudes. Canopus, the second-brightest star, also looms high, more toward the south. Its far-southern declination (–53°) makes it invisible from midnorthern latitudes. This is a pity, as the Great Star of the South is a brilliant celestial beacon, separating the Milky Way clouds of Vela and Puppis from the much more sparsely populated regions of Pictor and Dorado.

Canopus marks the rudder of the great ship Argo Navis, which is now composed of four southern constellations: Carina, the Keel; Vela, the Sails; Puppis, the Stern; and Pyxis, the Compass. Canopus is at the extreme western corner of Carina. This watery connection is appropriate for Canopus since it is named after the chief pilot of Menelaus's fleet, which destroyed Troy in 1183 B.C. Alas, Canopus died on the return voyage from the war, but he was honored when his grateful master named a city after him. The city of Canopus, long since ruined, was near Alexandria, the center of ancient scholarship that became the home of the great astronomer Ptolemy over a thousand years later. From there the star Canopus culminates today about 7½° above the southern horizon. It rose higher in ancient times because of precession.

INTERESTING OBJECTS

Object	Constellation	Type	Right Ascension	Dec.	Size/Sep./ Period	Magnitude
Hyades	Taurus	OC	04^h 27^m	+16° 00′	6°	0.5
Large Magellanic Cloud	Dorado	G	05^h 23^m.6	–69° 45′	11° × 9°	0.6
Small Magellanic Cloud	Tucana	G	00^h 52^m.7	–72° 50′	5° × 3°	2.8
M1 (Crab Nebula)	Taurus	DN	05^h 34^m.5	+22° 01′	6′ × 4′	8
M41	Canis Major	OC	06^h 47^m.0	–20° 44′	38′	4.5
M42 (Orion Nebula)	Orion	DN	05^h 35^m.4	–05° 27′	66′ × 60′	4
M46	Puppis	OC	07^h 41^m.8	–14° 49′	27′	6.1
M50	Monoceros	OC	07^h 03^m.2	–08° 20′	16′	5.9
M79	Lepus	GC	05^h 46^m.7	–24° 33′	9′	8.0
M93	Puppis	OC	07^h 44^m.6	–23° 52′	22′	6
NGC 2244	Monoceros	OC	06^h 32^m.4	+04° 52′	24′	4.8
NGC 2477	Puppis	OC	07^h 52^m.3	–38° 33′	27′	5.8
NGC 2516	Carina	OC	07^h 58^m.3	–60° 52′	30′	3.8
30 Doradus (Tarantula Nebula, NGC 2070)	Dorado	DN	05^h 38^m.7	–69° 06′	40′ × 25′	5
α Orionis (Betelgeuse)	Orion	VS	05^h 55^m.2	+07° 24′	5.8 years	0.4–1.3

DN = diffuse nebula; DS = double or multiple star; DkN = dark nebula; G = galaxy; GC = globular cluster; OC = open cluster; PN = planetary nebula; VS = variable star

Canopus was regarded as an important star back then, especially by the desert dwellers of the Mideast. As early as 6400 B.C., temples were constructed on the Nile in its honor; when Canopus shone into them it heralded the autumnal equinox. This star is also said to have given rise to many of the proverbs, stories, and superstitions of the early Arabs, who knew it by the name Suhail. (This name was also applied to the star Lambda Velorum). One tale relates how Suhail went a-wooing Al Jauzah (Orion's stars). Al Jauzah not only refused him, but very unceremoniously kicked him into the southern heavens.

South of the equator, where Canopus rises much higher in the sky, Polynesian mariners knew the star as Autahi (or Atutahi), one of the children of Rangi, the sky god. To the Polynesians Canopus was an important navigational aid, used during long ocean voyages by the one or two expert stargazers carried on each canoe. During the voyage of the canoe Takitimu from eastern Polynesia, the ancestors of the Maori people of New Zealand, were said to have used the stars Atutahi (Canopus), Puanga (Rigel), and Takurua (Sirius) to successfully guide them to Aotearoa (New Zealand).

M93 in Puppis

M41 in Canis Major

From their new country the Maori greeted the evening reappearance of Canopus in the southern sky with celebration, as it marked the season when kumara (sweet potatoes) were to be planted. The aspect of Canopus was also important in foretelling the coming season's weather, for tradition stated that if the star's rays shone toward the south, rain and snow would follow, but if toward the north, the season would be mild and warm.

Canopus's placement on the outskirts of the Milky Way made it a *tapu,* or sacred star, to the Maori. Perhaps this was because a very sacred person in Maori society was the tohunga (priest), who dwelled separately from other members of his tribe. However, the origin of Canopus's aloofness is also explained in one Maori myth concerning Tane, son of Rangi, the Sky Father, and Papa, the Earth Mother. Tane went star-gathering one day, heaping the shining ones into the basket of the Milky Way. But poor Canopus was left hanging outside while the laden basket was placed in the canoe Te Waka-o-Tama-rereti — which today we know as the tail of the Scorpion.

In spite of its lonely position, the Maori people did connect Canopus with other bright stars in the sky. The wide angle from Canopus through Sirius to Procyon, for example, formed the constellation Maan, said to represent a bird with a broken wing. In February Maan flies high in our sky — despite its handicap — throughout the evening.

— Graham Blow

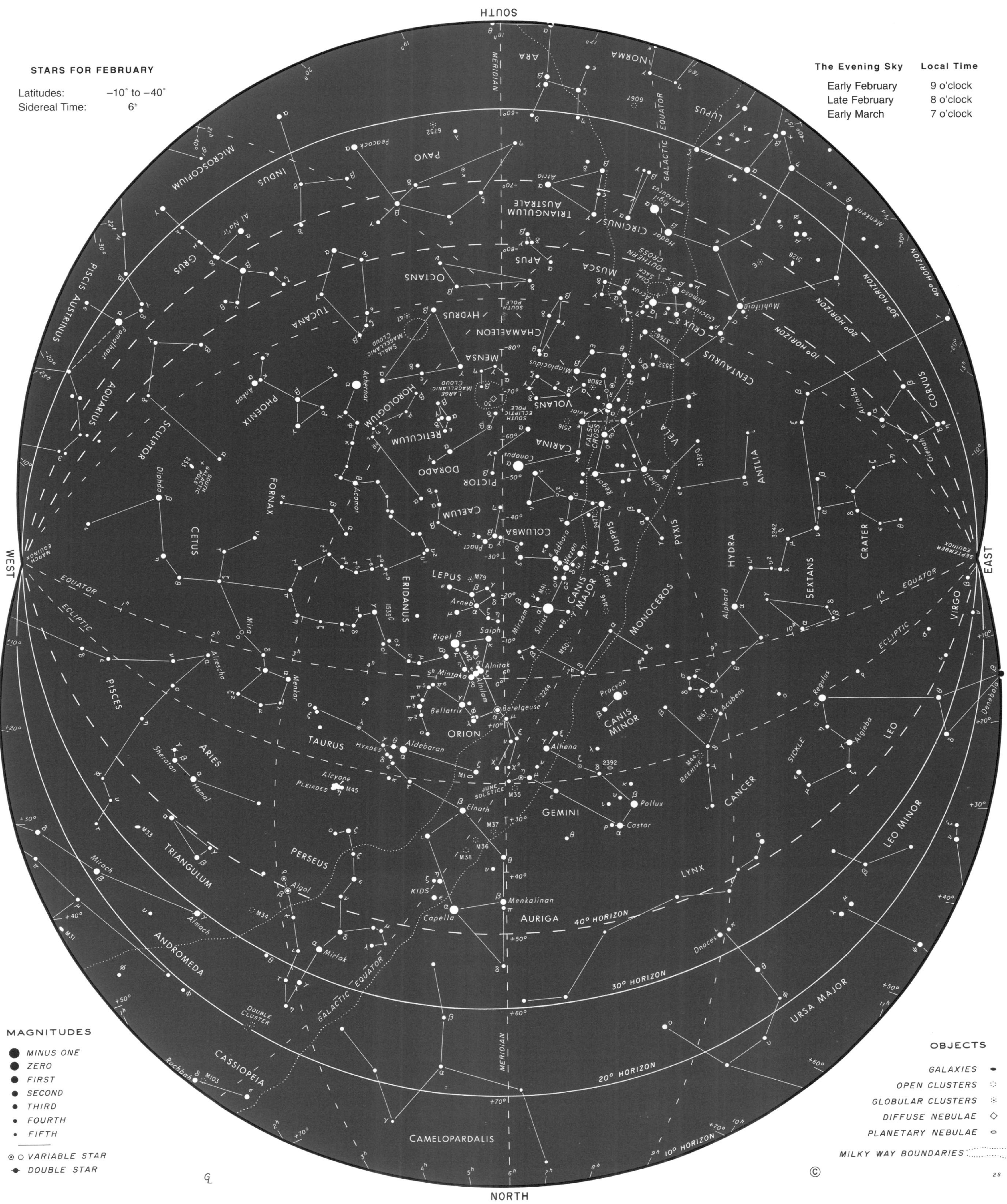
STARS FOR FEBRUARY
Latitudes: −10° to −40°
Sidereal Time: 6h
The Evening Sky | Local Time
Early February 9 o'clock
Late February 8 o'clock
Early March 7 o'clock
SOUTH
NORTH
EAST
WEST
MAGNITUDES
MINUS ONE
ZERO
FIRST
SECOND
THIRD
FOURTH
FIFTH
VARIABLE STAR
DOUBLE STAR
OBJECTS
GALAXIES
OPEN CLUSTERS
GLOBULAR CLUSTERS
DIFFUSE NEBULAE
PLANETARY NEBULAE
MILKY WAY BOUNDARIES
ORION
TAURUS
GEMINI
AURIGA
CANIS MAJOR
CANIS MINOR
MONOCEROS
LEPUS
ERIDANUS
CETUS
PISCES
ARIES
TRIANGULUM
PERSEUS
ANDROMEDA
CASSIOPEIA
CAMELOPARDALIS
LYNX
CANCER
LEO
LEO MINOR
URSA MAJOR
HYDRA
SEXTANS
CRATER
CORVUS
VIRGO
ANTLIA
PYXIS
PUPPIS
VELA
CARINA
CENTAURUS
CRUX
COLUMBA
CAELUM
PICTOR
DORADO
RETICULUM
HOROLOGIUM
FORNAX
SCULPTOR
PHOENIX
AQUARIUS
PISCES AUSTRINUS
GRUS
MICROSCOPIUM
INDUS
TUCANA
HYDRUS
OCTANS
MENSA
CHAMAELEON
VOLANS
APUS
MUSCA
CIRCINUS
TRIANGULUM AUSTRALE
PAVO
ARA
NORMA
LUPUS
EQUATOR
ECLIPTIC
GALACTIC EQUATOR
MERIDIAN
40° HORIZON
30° HORIZON
20° HORIZON
10° HORIZON

MARCH

THE TREASURES OF ARGO

RIDING prominently in the northwest this month is the constellation many regard as the most spectacular — Orion. But high overhead, hugging the meridian, is another grouping that once held that status: the ancient ship Argo Navis, whose bright stars and variety of interesting celestial objects are hard to match anywhere.

Today this huge former constellation is broken up into four parts: Puppis, the Stern; Vela, the Sails; Carina, the Keel; and Pyxis, the Mariner's Compass (which was once named Malus, the Mast). This heavenly vessel has accumulated lore and mythology from various early peoples, but it is most often associated with the Greek legend of Jason and his crew of Argonauts, who set out in quest of the Golden Fleece. Only hints of Argo's glories can be seen from the United States; it sails high in the sky only when viewed from south of the equator.

Most of Argo's bright stars are members of a swath of blue giants that stretches along the southern Milky Way from Orion to Scorpius. But there are exceptions. Look at Lambda (λ) Velorum, or Suhail, the northern member of the prominent 2nd-magnitude quadrilateral marking Vela. Suhail's orange color (spectral type *K5*) really stands out from all the blue-white beacons in this part of the sky. In fact, to appreciate the contrast, look at the star to its left in the quadrilateral — Gamma (γ), or Regor. This is the brightest of the rare, ultra hot Wolf-Rayet stars, type *W*, which come even before *O* in the spectral sequence.

As a matter of fact, you don't have to go far to find an *O* star here. Almost directly north of Gamma Velorum is Zeta (ζ) Puppis, and, speaking of coincidences, it is nearly the brightest of *its* spectral class. Both stars are unusually blue.

Have a good look at Gamma Velorum with binoculars; it's a splendid low-power double with a 4th-magnitude companion 41 arc seconds to the southwest. As we move down to the southernmost reaches of our ship, toward Carina, we encounter a plethora of deep-sky treats. Most common here are open, or galactic, clusters, but nebulae are also quite numerous.

As you explore this area, start off with the bright but loose open cluster IC 2602 (not labeled on the map) surrounding 3rd-magnitude Theta (θ) Carinae. Once, while en route from Australia to Indonesia aboard a 747 full of eclipse chasers, I chatted with some Australian amateurs who told me they refer to IC 2602 as the "Southern Pleiades." Having seen it just a few days previously, I could appreciate why. It does suggest a slightly fainter Pleiades and, like the Seven Sisters, has a central 3rd-magnitude star.

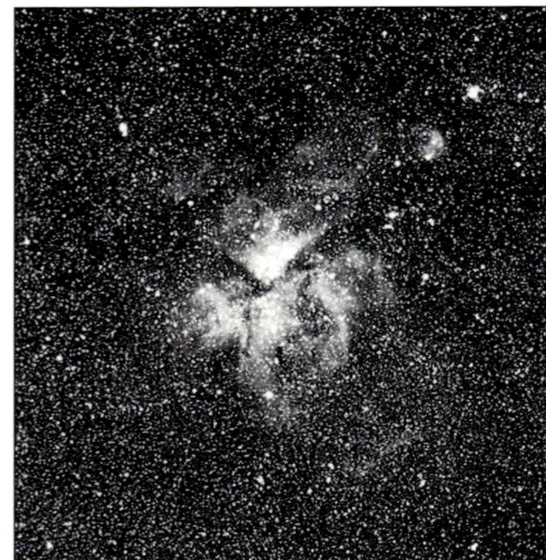

Eta Carinae Nebula

M46 in Puppis

North of IC 2602 is that beautiful and mysterious object Eta (η) Carinae, a strange type of recurring nova. In 1843 it grew almost as bright as Sirius, and some astronomers consider it a likely candidate for a future supernova. Eta isn't much to look at now; the real drawing card is the gorgeous nebulosity that surrounds it, NGC 3372, known as the Keyhole or Eta Carinae Nebula. It's easily spotted with the naked eye as a diffuse patch against the Milky Way background. To see this object really well, you must get far away from built-up "conurbations" (as the Australians like to call metropolitan areas) on a moonless night.

Two other "don't miss" objects are the bright open clusters NGC 2516, southwest of Epsilon (ε) Carinae, and IC 2391 (not labeled on the map), grouped around 4th-magnitude Omicron (o) Velorum. Both clusters are naked-eye objects.

Any northerner who travels to far southern lands should certainly plan to make skywatching a major part of the trip. And I don't mean the Caribbean or Hawaii. Go far south of the equator, where these objects are high up and unaffected by horizon haze and dimming. You'll come home spoiled!

— George Lovi

NGC 2392 in Gemini

INTERESTING OBJECTS

Object	Constellation	Type	Right Ascension	Dec.	Size/Sep./ Period	Magnitude
Large Magellanic Cloud	Dorado	G	05ʰ 23ᵐ.6	−69° 45′	11° × 9°	0.6
Small Magellanic Cloud	Tucana	G	00ʰ 52ᵐ.7	−72° 50′	5° × 3°	2.8
M41	Canis Major	OC	06ʰ 47ᵐ.0	−20° 44′	38′	4.5
M44 (Beehive Cluster)	Cancer	OC	08ʰ 40ᵐ.1	+19° 59′	95′	3.1
M46	Puppis	OC	07ʰ 41ᵐ.8	−14° 49′	27′	6.1
M50	Monoceros	OC	07ʰ 03ᵐ.2	−08° 20′	16′	5.9
M67	Cancer	OC	08ʰ 50ᵐ.4	+11° 49′	30′	6.9
M93	Puppis	OC	07ʰ 44ᵐ.6	−23° 52′	22′	6
NGC 2392 (Eskimo Nebula)	Gemini	PN	07ʰ 29ᵐ.2	+20° 55′	1′	10
NGC 2477	Puppis	OC	07ʰ 52ᵐ.3	−38° 33′	27′	5.8
NGC 2516	Carina	OC	07ʰ 58ᵐ.3	−60° 52′	30′	3.8
NGC 2808	Carina	GC	09ʰ 12ᵐ.0	−64° 52′	14′	6.3
NGC 3532	Carina	OC	11ʰ 06ᵐ.4	−58° 40′	55′	3.0
η Carinae Nebula (NGC 3372)	Carina	DN	10ʰ 43ᵐ.8	−59° 52′	2°	3
l Carinae	Carina	VS	09ʰ 45ᵐ.2	−62° 30′	35.5 days	3.3–4.2
30 Doradus (Tarantula Nebula, NGC 2070)	Dorado	DN	05ʰ 38ᵐ.7	−69° 06′	40′ × 25′	5

DN = diffuse nebula; DS = double or multiple star; DkN = dark nebula; G = galaxy; GC = globular cluster; OC = open cluster; PN = planetary nebula; VS = variable star

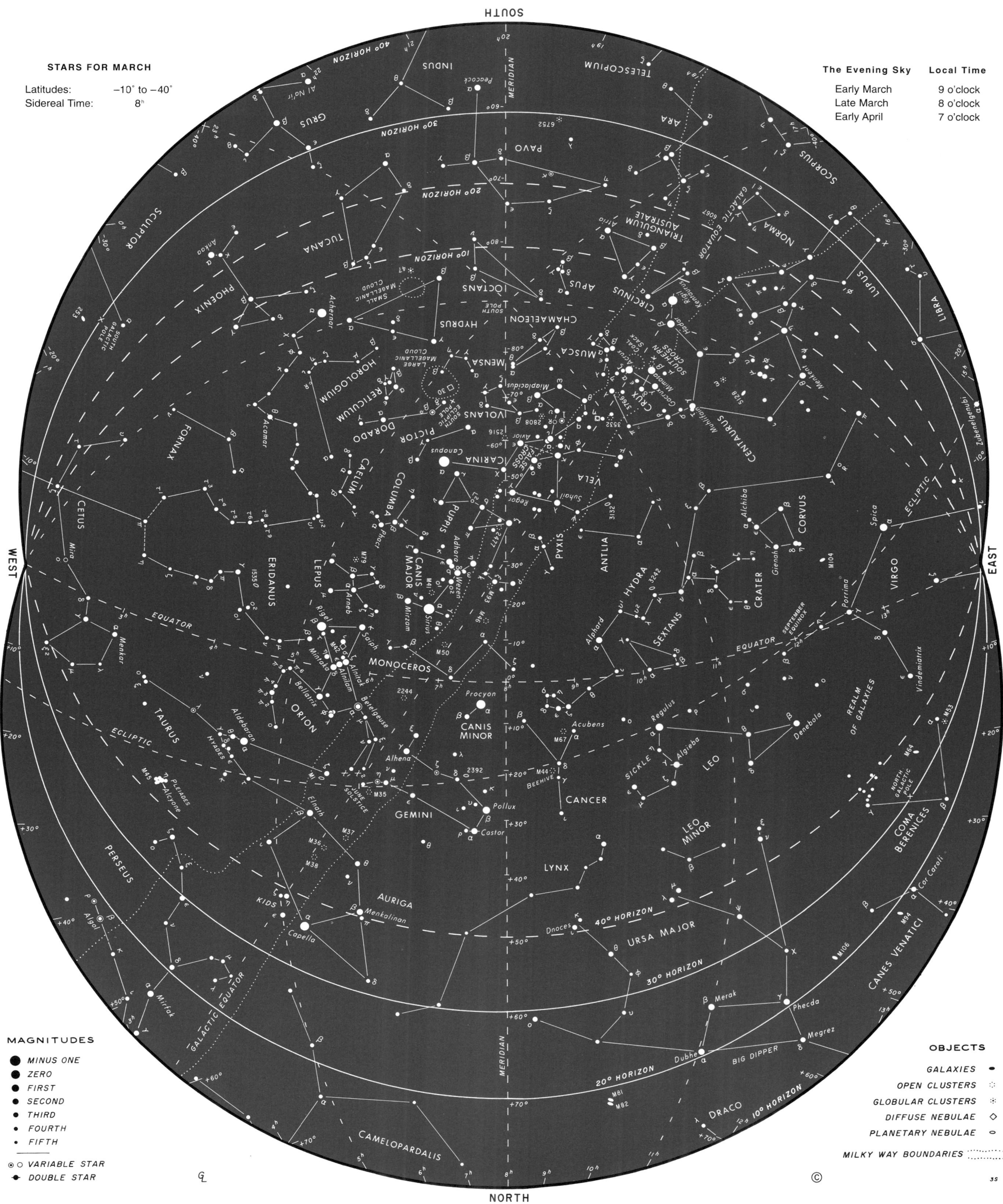
STARS FOR MARCH
Latitudes: −10° to −40°
Sidereal Time: 8h
The Evening Sky
Local Time
Early March 9 o'clock
Late March 8 o'clock
Early April 7 o'clock
SOUTH
NORTH
WEST
EAST
MERIDIAN
EQUATOR
ECLIPTIC
GALACTIC EQUATOR
10° HORIZON
20° HORIZON
30° HORIZON
40° HORIZON
INDUS
TELESCOPIUM
GRUS
PAVO
ARA
SCORPIUS
SCULPTOR
TUCANA
TRIANGULUM AUSTRALE
NORMA
PHOENIX
SMALL MAGELLANIC CLOUD
OCTANS
APUS
CIRCINUS
LUPUS
LIBRA
HYDRUS
SOUTH POLE
CHAMAELEON
HOROLOGIUM
LARGE MAGELLANIC CLOUD
MENSA
MUSCA
SOUTHERN CROSS
CRUX
RETICULUM
DORADO
VOLANS
CENTAURUS
FORNAX
PICTOR
CARINA
FALSE CROSS
VELA
CAELUM
COLUMBA
PUPPIS
CORVUS
CETUS
ERIDANUS
LEPUS
CANIS MAJOR
PYXIS
ANTLIA
CRATER
VIRGO
HYDRA
SEXTANS
MONOCEROS
ORION
TAURUS
CANIS MINOR
CANCER
LEO
REALM OF GALAXIES
SEPTEMBER EQUINOX
JUNE SOLSTICE
GEMINI
SICKLE
COMA BERENICES
NORTH GALACTIC POLE
PERSEUS
LEO MINOR
LYNX
AURIGA
KIDS
URSA MAJOR
CANES VENATICI
BIG DIPPER
DRACO
CAMELOPARDALIS
Achernar
Canopus
Sirius
Rigel
Betelgeuse
Procyon
Aldebaran
Pollux
Castor
Regulus
Spica
Capella
Alphard
Denebola
Algol
Mirfak
Dubhe
Merak
Phecda
Megrez
Cor Caroli
Menkalinan
Elnath
Alhena
Hyades
Pleiades
Alcyone
Beehive
Acubens
Algieba
Porrima
Vindemiatrix
Menkar
Mira
Acamar
Ankaa
Al Na'ir
Peacock
Atria
Hadar
Miaplacidus
Avior
Suhail
Regor
Adhara
Wezen
Mirzam
Saiph
Mintaka
Alnilam
Alnitak
Bellatrix
Arneb
Phact
Alchiba
Gienah
Zubenelgenubi
Menkent
Muliphein
Dnoces
MAGNITUDES
MINUS ONE
ZERO
FIRST
SECOND
THIRD
FOURTH
FIFTH
VARIABLE STAR
DOUBLE STAR
OBJECTS
GALAXIES
OPEN CLUSTERS
GLOBULAR CLUSTERS
DIFFUSE NEBULAE
PLANETARY NEBULAE
MILKY WAY BOUNDARIES

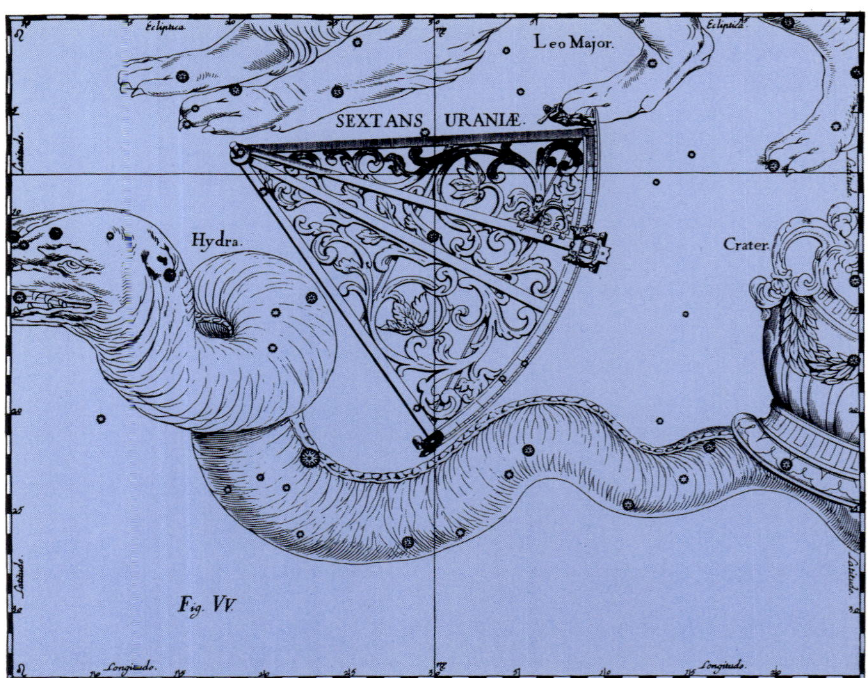

The Crux Connection

THE brilliant celestial sights of the ancient constellation Argo Navis were the subject of last month's discussion. But the spectacles of the far southern sky don't end there. Just east of Carina, and right above the southern horizon for 20° latitude, is the region around Crux, the Southern Cross. If you were to ask what is the most stunning part of the sky from a naked-eye standpoint, I would certainly say this is it.

Where else do you find four 1st-magnitude stars grouped within 16° of each other? Two belong to Crux, and the other two — the "guardian" stars of the cross — are in Centaurus. The brightest, Alpha Centauri, is actually zero magnitude. These four luminaries occupy a smaller sky area than the main stars of Orion, the usual candidate for the most spectacular part of the sky. Their visual impact is greatest in deep-blue twilight; they formed an indescribable sight when I saw them that way over the skyline of Melbourne, Australia!

To see Crux and company really well you need to be south of the equator. However, they are very attractive even from as far north as the Caribbean or Hawaii, especially if viewed over the ocean at dusk.

The stars forming our present-day Southern Cross were originally part of huge, neighboring Centaurus. The invention of Crux is usually ascribed to the 17th-century Frenchman Augustin Royer. He is said to have been influenced by various Southern Hemisphere explorers who saw in these stars a symbol of their religion, which served as a celestial amulet to protect them from the numerous dangers in that unknown part of the world. Crux is also the sole remnant of the 17th-century movement to "depaganize" the constellations. An example of this effort was the atlas of Julius Schiller, who converted ancient constellations into biblical figures.

Interesting Objects

Object	*Constellation*	*Type*	*Right Ascension*	*Dec.*	*Size/Sep./ Period*	*Magnitude*
Coalsack Nebula	Crux	DkN	12^h 53^m	–63° 03′	7° × 5°	
Large Magellanic Cloud	Dorado	G	05^h 23^m.6	–69° 45′	11° × 9°	0.6
M44 (Beehive Cluster)	Cancer	OC	08^h 40^m.1	+19° 59′	95′	3.1
NGC 2516	Carina	OC	07^h 58^m.3	–60° 52′	30′	3.8
NGC 3132	Vela	PN	10^h 07^m.0	–40° 26′	0.8′	8
NGC 3532	Carina	OC	11^h 06^m.4	–58° 40′	55′	3.0
NGC 3766	Centaurus	OC	11^h 36^m.1	–61° 37′	12′	5.3
η Carinae Nebula (NGC 3372)	Carina	DN	10^h 43^m.8	–59° 52′	2°	3
l Carinae	Carina	VS	09^h 45^m.2	–62° 30′	35.5 days	3.3–4.2
α Crucis (Acrux)	Crux	DS	12^h 26^m.6	–63° 06′	AB 4″.0 AC 90″.1	1.4, 1.9 4.9
κ Crucis (Jewel Box Cluster, NGC 4755)	Crux	OC	12^h 53^m.6	–60° 20′	10′	4.2
30 Doradus (Tarantula Nebula, NGC 2070)	Dorado	DN	05^h 38^m.7	–69° 06′	40′ × 25′	5

DN = diffuse nebula; DS = double or multiple star; DkN = dark nebula; G = galaxy; GC = globular cluster; OC = open cluster; PN = planetary nebula; VS = variable star

April

Even before Crux became a separate constellation, explorers and travelers repeatedly referred to the "four stars" seen from the Southern Hemisphere. Among those doing so was the Italian navigator Amerigo Vespucci, for whom the Americas are named. Even Dante mentions "the four stars of the south." I wonder, however, if those who spoke of the "four stars" really meant not Crux but the four 1st-magnitude luminaries described earlier. They really hit you in the eye!

An interesting tradition pertaining to Crux is described by Peter Lum in his book *The Stars in Our Heaven.* According to Lum, the Persians observed a special feast in honor of these stars, which, owing to the sky's precession, were visible from Persia's northern latitude many centuries ago. With the passage of time and the inexorable precessional creep, Crux worked its way to the southern horizon and below. So the celebrants replaced Crux with Delphinus, another somewhat diamond-shaped asterism but hardly a worthy substitute in size and splendor.

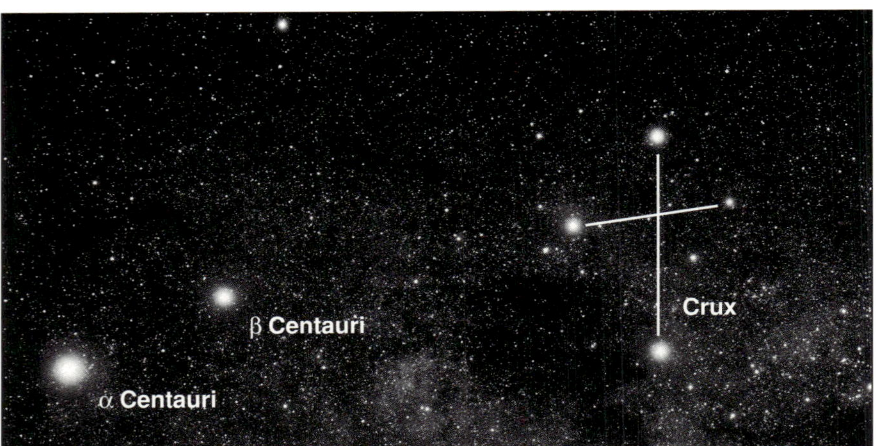

Alpha and Beta Centauri and the Southern Cross

Actually, Crux most resembles a lopsided kite. There's a debate over why these stars have been made into a cross when there is no prominent star marking the intersection of the arms. In the *Field Guide to the Stars and Planets,* originally by Donald Menzel and completely reworked by Jay M. Pasachoff, Crux is deliberately portrayed as a kite.

However, there *is* a naked-eye star marking the arms' intersection: BS 4768, although it is only 5th magnitude. If we add this star to the pattern, along with Epsilon (with which Crux is usually portrayed), we find that the Southern Cross replicates itself twice fairly well by rotating to the right and shrinking, as shown in the illustration below.

Crux also has deep-sky treats. The leader is the incomparable open cluster Kappa (κ) Crucis, or NGC 4755. It is just southeast of Beta (β) Crucis. Here is a case of slightly confused nomenclature — Kappa is actually a blue star of magnitude 5.95 inside the cluster, but in popular usage the entire stellar conglomeration is now called Kappa Crucis, which is how it's labeled on our all-sky chart.

But the star Kappa is not the real grabber in the cluster. Tucked among all those blue gems is a solitary red star, SAO 252073. Here is another excellent place to observe star colors by contrast. John Herschel's description of this cluster gave rise to the "Jewel Box" nickname by which it's so well known today. And that's just what it suggests: a clutch of diamonds with a single ruby in their midst.

— George Lovi

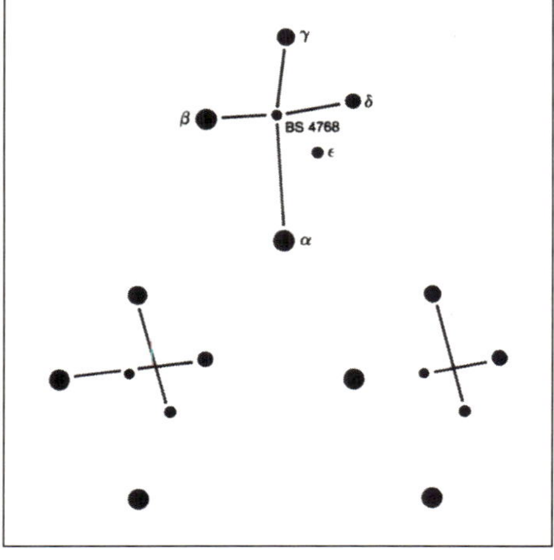

Multiple crosses can be formed by joining stars in the constellation Crux.

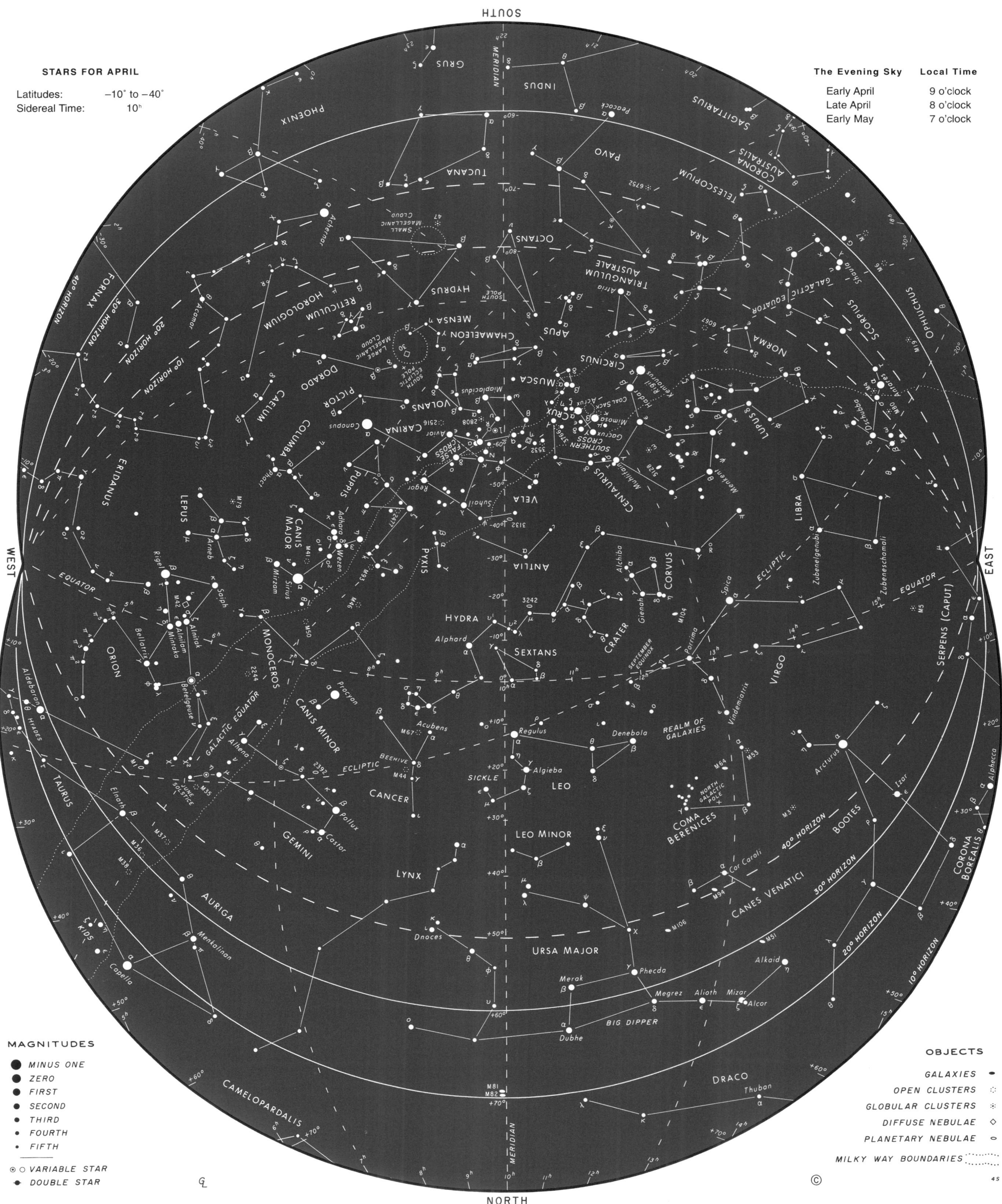
STARS FOR APRIL
Latitudes: −10° to −40°
Sidereal Time: 10h
The Evening Sky / Local Time
Early April 9 o'clock
Late April 8 o'clock
Early May 7 o'clock
SOUTH
NORTH
WEST
EAST
MAGNITUDES
MINUS ONE
ZERO
FIRST
SECOND
THIRD
FOURTH
FIFTH
VARIABLE STAR
DOUBLE STAR
OBJECTS
GALAXIES
OPEN CLUSTERS
GLOBULAR CLUSTERS
DIFFUSE NEBULAE
PLANETARY NEBULAE
MILKY WAY BOUNDARIES
MERIDIAN
EQUATOR
ECLIPTIC
GALACTIC EQUATOR
HYDRA
SEXTANS
LEO
LEO MINOR
URSA MAJOR
BIG DIPPER
CANCER
GEMINI
LYNX
AURIGA
TAURUS
ORION
MONOCEROS
CANIS MINOR
CANIS MAJOR
LEPUS
ERIDANUS
CAELUM
COLUMBA
PUPPIS
VELA
PYXIS
ANTLIA
CRATER
CORVUS
VIRGO
LIBRA
CENTAURUS
LUPUS
NORMA
SCORPIUS
OPHIUCHUS
SERPENS (CAPUT)
BOOTES
CORONA BOREALIS
COMA BERENICES
CANES VENATICI
DRACO
CAMELOPARDALIS
CARINA
VOLANS
PICTOR
DORADO
MENSA
CHAMAELEON
MUSCA
CRUX
CIRCINUS
APUS
TRIANGULUM AUSTRALE
ARA
OCTANS
HYDRUS
RETICULUM
HOROLOGIUM
FORNAX
PHOENIX
TUCANA
PAVO
INDUS
GRUS
TELESCOPIUM
CORONA AUSTRALIS
SAGITTARIUS
REALM OF GALAXIES
SEPTEMBER EQUINOX
JUNE SOLSTICE
NORTH GALACTIC POLE
SOUTH ECLIPTIC POLE
SOUTH POLE
SMALL MAGELLANIC CLOUD
LARGE MAGELLANIC CLOUD
SOUTHERN CROSS
FALSE CROSS
COAL SACK
HYADES
KIDS
SICKLE
BEEHIVE
40° HORIZON
30° HORIZON
20° HORIZON
10° HORIZON

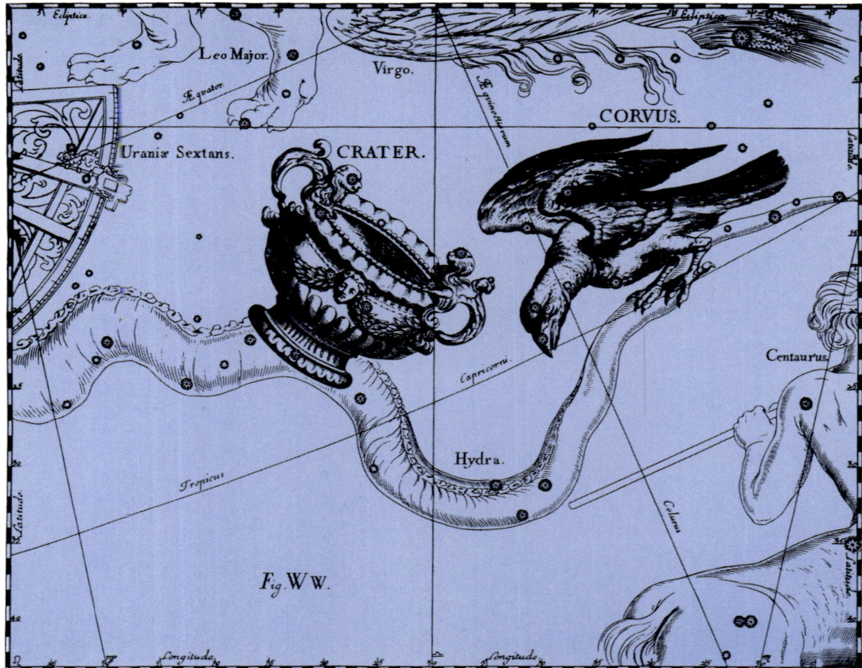

MAY

Crosses True and False

MOST southern observers have no trouble recognizing Crux, the Southern Cross, but visitors from north of the equator sometimes get confused by another group of stars that looks quite similar — the False Cross. Located high in the southwest after sunset, it is formed by the stars Iota (ι) and Epsilon (ε) Carinae and Kappa (κ) and Delta (δ) Velorum. Both the Southern Cross and the False Cross are shaped like kites, and they are even oriented similarly — but there are a few important differences. The most obvious is that the False Cross lacks two bright attendant stars "pointing" to it the way Alpha and Beta Centauri accompany Crux. Neither does the False Cross have any 1st-magnitude stars, and its long axis is about 50 percent longer than the exactly 6° separating Alpha (α) and Gamma (γ) Crucis (which, incidentally, provide a most useful reference for measuring other short distances in this part of the sky).

It's important to avoid confusing Crux and the False Cross, because anyone who can recognize the Southern Cross has a sure-fire method of finding due south. Simply extend the long arm of Crux through its base 4½ times and you're within 3° of the south celestial pole.

The Southern Cross and False Cross form bookends to one of the most interesting areas of the southern Milky Way. This includes the great nebula around Eta (η) Carinae and a whole raft of clusters and variable stars. Many years ago when I first became interested in astronomy, this was the part of the sky I found the most fascinating.

One object I examined frequently lies just southeast of the False Cross. In fact, a line extended from Delta Velorum through Iota Carinae takes you right to it. This is the star l Carinae (that's "el," not "one"), a variable of the Cepheid type, and among the brightest of its class. Because I lived in midsouthern latitudes I was denied a view of the prototype Cepheid variable Delta Cephei — but l Carinae was a most acceptable substitute.

Watching the variations of l Carinae was one of my first projects in observational astronomy. At about 4th magnitude it was very easy to find, so every few nights I would take my father's binoculars outside to observe it. Using a chart from the Royal Astronomical Society of New Zealand's Variable Star Section (the Southern Hemisphere's equivalent of the American Association of Variable Star Observers), I would compare it against the known brightnesses of nearby stars. With great enthusiasm I would then hurry inside to plot my observation on a graph (which I still have today, over 20 years later). After about a month I had enough observations to see that l Carinae had brightened from about magnitude 4.2 to 3.3 and then declined again. My graph also showed that its period must be around 35 days. What a thrill to realize that the universe was putting on a performance before my very eyes!

M64 in Coma Berenices

Coalsack Nebula in Crux

This region of sky also contains a star cluster that is one of the most underrated in the Southern Hemisphere. Extend a line from Kappa Velorum through Epsilon Carinae and you come to NGC 2516, a beautiful open cluster easily visible to the naked eye. This cluster has been compared in beauty with Praesepe, the Beehive Cluster in Cancer. Binoculars show its densest region tucked neatly into the southeastern side of a slightly elongated hexagon of 5th- and 6th-magnitude stars. The cluster also has several nice double stars and a particularly red star on its southeastern edge. Although NGC 2516 is much brighter than the better-known M6, it doesn't seem to be mentioned nearly as often — probably because its far southern declination makes it invisible to most Northern Hemisphere observers.

Before we leave this region of celestial crosses, one further confusing asterism needs mention. Immediately south of the False Cross the stars Beta (β), Upsilon (υ), Theta (θ), and Omega (ω) Carinae form what is sometimes known as the Diamond Cross. This grouping is perhaps the most true to its name, for with its equal sides it really does look like a perfect diamond in the sky. However, most of its hot *O* and *B* stars are below 2nd magnitude, which makes it a little less obvious to the eye and helps in avoiding confusion with its adjacent, more familiar cousins, the Southern Cross and False Cross.

— Graham Blow

Interesting Objects

Object	Constellation	Type	Right Ascension	Dec.	Size/Sep./ Period	Magnitude
Coalsack Nebula	Crux	DkN	12^h 53^m	−63° 03′	7°×5°	
Coma Cluster	Coma Ber.	OC	12^h 25^m	+26° 00′	5°	1.8
M53	Coma Ber.	GC	13^h 12^m.9	+18° 10′	13′	7.7
M64 (Black-eye Galaxy)	Coma Ber.	G	12^h 56^m.7	+21° 41′	9′ × 5′	8.5
M104 (Sombrero Galaxy)	Virgo	G	12^h 40^m.0	−11° 37′	9′ × 4′	8.3
NGC 2808	Carina	GC	09^h 12^m.0	−64° 52′	13′.8	6.3
NGC 3132	Vela	PN	10^h 07^m.0	−40° 26′	0′.8	8
NGC 3242 (Ghost of Jupiter)	Hydra	PN	10^h 24^m.8	−18° 38′	20′.8	9
NGC 3532	Carina	OC	11^h 06^m.4	−58° 40′	55′	3.0
NGC 3766	Centaurus	OC	11^h 36^m.1	−61° 37′	12′	5.3
η Carinae Nebula (NGC 3372)	Carina	DN	10^h 43^m.8	−59° 52′	2°	3
l Carinae	Carina	VS	09^h 45^m.2	−62° 30′	35.5 days	3.28–4.18
α Centauri	Centaurus	DS	14^h 39^m.6	−60° 50′	14″	0.0, 1.2
ω Centauri (NGC 5139)	Centaurus	GC	13^h 26^m.8	−47° 29′	36′	3.7
α Crucis (Acrux)	Crux	DS	12^h 26^m.6	−63° 06′	AB 4″.0 AC 90″.1	1.4, 1.9 4.9
κ Crucis (Jewel Box Cluster, NGC 4755)	Crux	OC	12^h 53^m.6	−60° 20′	10′	4.2

DN = diffuse nebula; DS = double or multiple star; DkN = dark nebula; G = galaxy; GC = globular cluster; OC = open cluster; PN = planetary nebula; VS = variable star

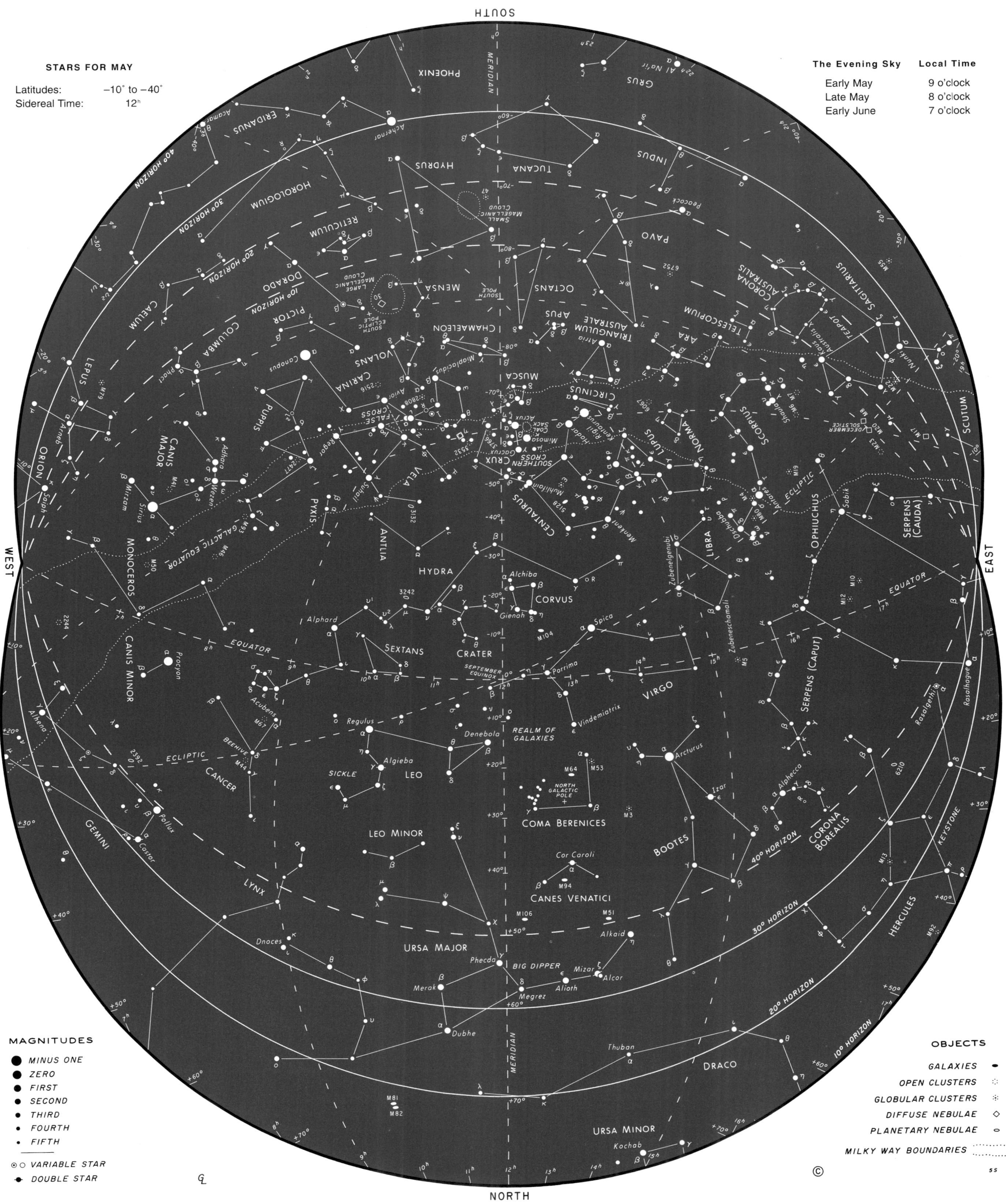
STARS FOR MAY
Latitudes: −10° to −40°
Sidereal Time: 12h
The Evening Sky | Local Time
Early May 9 o'clock
Late May 8 o'clock
Early June 7 o'clock
SOUTH
NORTH
WEST
EAST
MERIDIAN
PHOENIX
GRUS
ERIDANUS
HYDRUS
TUCANA
INDUS
HOROLOGIUM
RETICULUM
PAVO
DORADO
MENSA
OCTANS
CORONA AUSTRALIS
SAGITTARIUS
CAELUM
PICTOR
COLUMBA
CHAMAELEON
APUS
TRIANGULUM AUSTRALE
TELESCOPIUM
ARA
LEPUS
VOLANS
CARINA
MUSCA
CIRCINUS
NORMA
SCORPIUS
PUPPIS
FALSE CROSS
LUPUS
SCUTUM
ORION
CANIS MAJOR
VELA
CRUX
SOUTHERN CROSS
CENTAURUS
OPHIUCHUS
SERPENS (CAUDA)
PYXIS
ANTLIA
LIBRA
MONOCEROS
HYDRA
CORVUS
GALACTIC EQUATOR
EQUATOR
SEXTANS
CRATER
CANIS MINOR
SERPENS (CAPUT)
VIRGO
REALM OF GALAXIES
ECLIPTIC
CANCER
LEO
SICKLE
COMA BERENICES
GEMINI
LEO MINOR
BOOTES
CORONA BOREALIS
KEYSTONE
CANES VENATICI
LYNX
HERCULES
URSA MAJOR
BIG DIPPER
DRACO
URSA MINOR
40° HORIZON
30° HORIZON
20° HORIZON
10° HORIZON
MAGNITUDES
MINUS ONE
ZERO
FIRST
SECOND
THIRD
FOURTH
FIFTH
VARIABLE STAR
DOUBLE STAR
OBJECTS
GALAXIES
OPEN CLUSTERS
GLOBULAR CLUSTERS
DIFFUSE NEBULAE
PLANETARY NEBULAE
MILKY WAY BOUNDARIES
©

JUNE

A Strange Case of Celestial Acne

WHEN Centaurus is mentioned at public star parties and observatory nights it is often in the context of its two brightest stars, Alpha (α) and Beta (β) Centauri. Not only does this pair form the well-known "Pointers" to the Southern Cross, but Alpha is a very interesting star system in its own right.

Alpha Centauri's greatest claim to fame is its proximity — a mere 4.3 light-years away, making it our solar system's closest neighbor. It is not just in our own backyard but right outside our door, astronomically speaking. However, I have always liked 19th-century astronomer John Herschel's description of its true distance: "To drop a pea at the end of every mile of a voyage on a limitless ocean to the nearest fixed star would require a fleet of 10,000 ships of 600 tons burthen, each starting with a full cargo of peas." This does help to maintain a sense of perspective!

I used the term *star system* above advisedly, because Alpha is also one of the most beautiful double stars in the sky. Its pale yellow and pale orange components differ in brightness by a little over one magnitude. Seen through a small telescope in deep twilight they look like two sparkling jewels against a turquoise background. Breathtaking!

Alpha Centauri's duplicity was discovered in 1689, when the two components were about 8 arc seconds apart. The pair has an orbital period of 80 years, during which their separation reaches as much as 21.7 arc seconds. This last happened in 1981. In 2038 the pair will be a real test for small telescopes at a mere 1.7 arc seconds.

Good things, however, seem to come in threes, and Alpha's third claim to fame is its distant and elusive third component. The 11th-magnitude Proxima is believed to be the closest of the three to us. It is separated from its 1st-magnitude companions by 2.2°, which makes it difficult to find unless you have a detailed chart or photograph.

Another thing that makes dim, orange Proxima really interesting is its activity as a flare star. On rare occasions these stars can brighten by several magnitudes in the space of a few seconds — enough to wake any dozing observer!

Proxima's flares are associated with giant versions of our own Sun's spots, so I like to think of these spotted stars as suffering from a form of celestial acne. Some stars are afflicted more than others, and results from the Hubble Space Telescope suggest that Proxima has an especially bad case — its spots may cover up to half its visible face.

With all this to see and do around Alpha Centauri alone, it is easy to forget that there's a whole constellation waiting to be explored. Centaurus — the "Horse-man Beast," as the Greeks called it — is the ninth largest, covering more than 1,000 square degrees. Eratosthenes (the librarian at Alexandria best known for his accurate measurement of the size of the Earth) associated Centaurus with Chiron, the son of Chronos and the ocean nymph Philyra. This Centaur was of mild disposition and noble look, quite unlike the threatening aspect of the sky's other Centaur, Sagittarius.

M53 in Coma Berenices

You can pick out the shape of Centaurus by starting from Alpha Centauri. This was known to the Arabs as Al Rijl al Kentaurus, the Centaur's Foot (whence the modern navigator's name of Rigil Kent). Hadar, the Arabic name for Beta Centauri (itself a close double) may denote the ground on which the Centaur is standing, though the name has occasionally been applied to Alpha as well. The Centaur's forelimb extends from Beta through Epsilon (ε) to Zeta (ζ), while the manly head lies farther north in the region of Theta (θ).

From Zeta, the hindquarters of our beast stretch across the glorious globular cluster Omega Centauri to the brilliant yellow (but close) double star Gamma (γ), and thence to Delta (δ). In binoculars the 2nd-magnitude Delta has beautiful 4th- and 6th-magnitude companions, all three with a blue-white hue.

To complete the Centaur draw a zig-zag line from Delta through Pi (π) to Lambda (λ); this forms the back leg. Just north of Lambda look out for the scattered cluster NGC 3766, which contains some pretty orange, yellow, white, and bluish stars. It's a lovely cluster for small telescopes.

Alpha and Beta Centauri often seem to steal the Centaur's show. But taking a lead from the Horse-man Beast and leaping across Crux to explore the outer reaches of this wonderful constellation can provide many satisfying rewards.

— Graham Blow

Interesting Objects

Object	Constellation	Type	Right Ascension	Dec.	Size/Sep./ Period	Magnitude
Coalsack Nebula	Crux	DkN	12^h 53^m	−63° 03′	7°×5°	
Coma Cluster	Coma Ber.	OC	12^h 25^m	+26° 00′	5°	1.8
M3	Canes Ven.	GC	13^h 42^m.2	+28° 23′	16′	6.4
M5	Serpens	GC	15^h 18^m.6	+02° 05′	17′	5.8
M53	Coma Ber.	GC	13^h 12^m.9	+18° 10′	13′	7.7
M64 (Black-eye Galaxy)	Coma Ber.	G	12^h 56^m.7	+21° 41′	9′ × 5′	8.5
M104 (Sombrero Galaxy)	Virgo	G	12^h 40^m.0	−11° 37′	9′ × 4′	8.3
NGC 3766	Centaurus	OC	11^h 36^m.1	−61° 37′	12′	5.3
NGC 5128	Centaurus	G	13^h 25^m.5	−43° 01′	18′.2	7.0
NGC 6067	Norma	OC	16^h 13^m.2	−54° 13′	13′	5.6
α Centauri	Centaurus	DS	14^h 39^m.6	−60° 50′	14″	0.0, 1.2
ω Centauri (NGC 5139)	Centaurus	GC	13^h 26^m.8	−47° 29′	36′	3.7
α Crucis (Acrux)	Crux	DS	12^h 26^m.6	−63° 06′	AB 4″.0 AC 90″.1	1.4, 1.9 4.9
κ Crucis (Jewel Box Cluster, NGC 4755)	Crux	OC	12^h 53^m.6	−60° 20′	10′	4.2
α Librae (Zubenelgenubi)	Libra	DS	14^h 50^m.9	−16° 02′	231″	2.8, 5.2
γ Virginis (Porrima)	Virgo	DS	12^h 41^m.7	−01° 27′	3″.0	3.5, 3.5

DN = diffuse nebula; DS = double or multiple star; DkN = dark nebula; G = galaxy; GC = globular cluster; OC = open cluster; PN = planetary nebula; VS = variable star

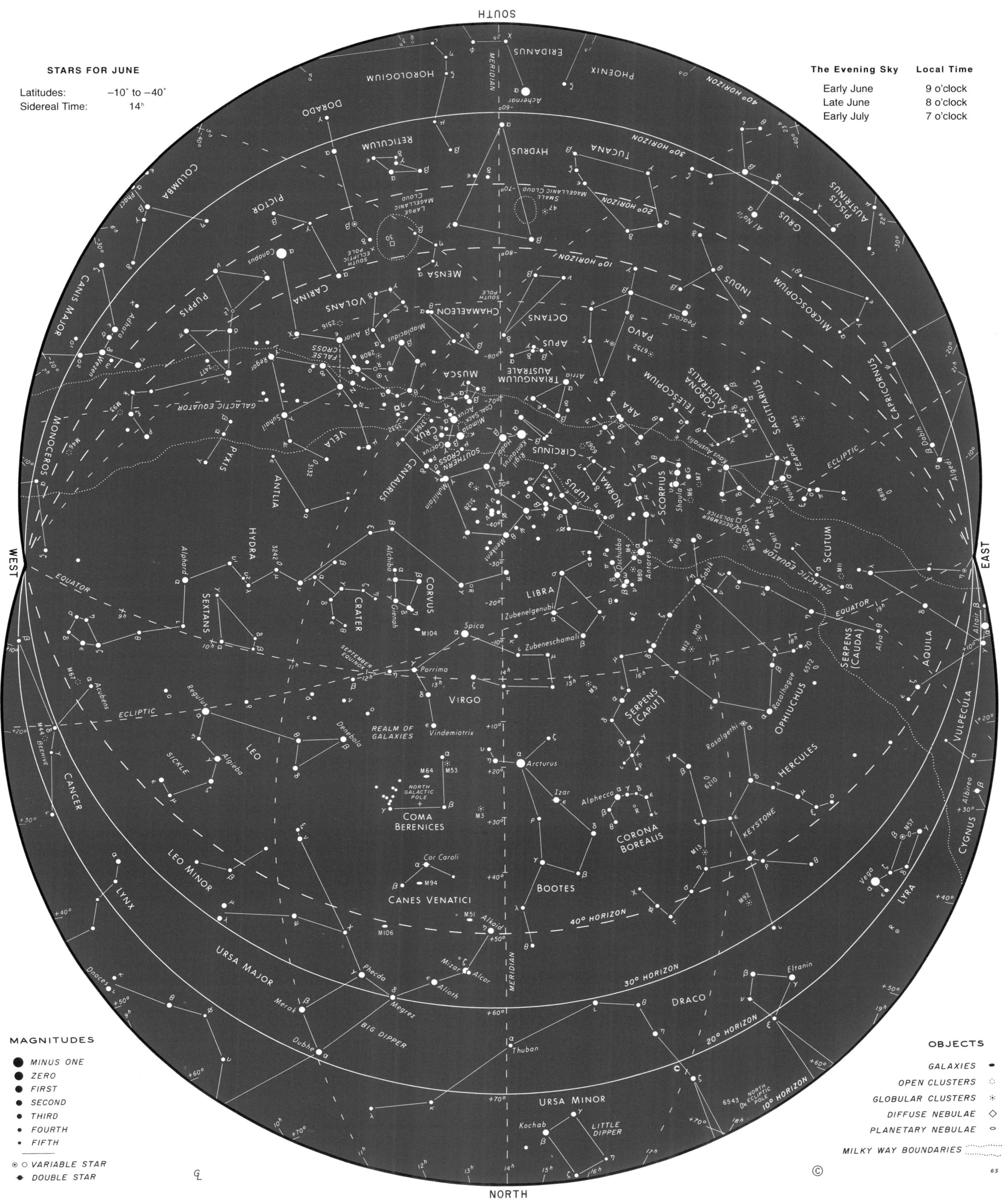

STARS FOR JUNE
Latitudes: −10° to −40°
Sidereal Time: 14h
The Evening Sky
Local Time
Early June 9 o'clock
Late June 8 o'clock
Early July 7 o'clock
SOUTH
NORTH
EAST
WEST
MAGNITUDES
MINUS ONE
ZERO
FIRST
SECOND
THIRD
FOURTH
FIFTH
VARIABLE STAR
DOUBLE STAR
OBJECTS
GALAXIES
OPEN CLUSTERS
GLOBULAR CLUSTERS
DIFFUSE NEBULAE
PLANETARY NEBULAE
MILKY WAY BOUNDARIES
ERIDANUS
PHOENIX
HOROLOGIUM
DORADO
RETICULUM
HYDRUS
TUCANA
COLUMBA
PICTOR
GRUS
PISCIS AUSTRINUS
MENSA
INDUS
MICROSCOPIUM
CANIS MAJOR
PUPPIS
CARINA
VOLANS
CHAMAELEON
OCTANS
PAVO
APUS
MUSCA
TRIANGULUM AUSTRALE
TELESCOPIUM
CORONA AUSTRALIS
SAGITTARIUS
CAPRICORNUS
MONOCEROS
PYXIS
VELA
CRUX
CENTAURUS
CIRCINUS
ARA
NORMA
LUPUS
SCORPIUS
ANTLIA
HYDRA
CRATER
CORVUS
LIBRA
SCUTUM
SEXTANS
VIRGO
SERPENS (CAUDA)
AQUILA
SERPENS (CAPUT)
OPHIUCHUS
LEO
CANCER
COMA BERENICES
HERCULES
CORONA BOREALIS
VULPECULA
CYGNUS
LEO MINOR
LYNX
CANES VENATICI
BOOTES
LYRA
URSA MAJOR
BIG DIPPER
DRACO
URSA MINOR
LITTLE DIPPER
REALM OF GALAXIES
NORTH GALACTIC POLE
SOUTH POLE
NORTH ECLIPTIC POLE
ECLIPTIC
EQUATOR
GALACTIC EQUATOR
MERIDIAN
10° HORIZON
20° HORIZON
30° HORIZON
40° HORIZON

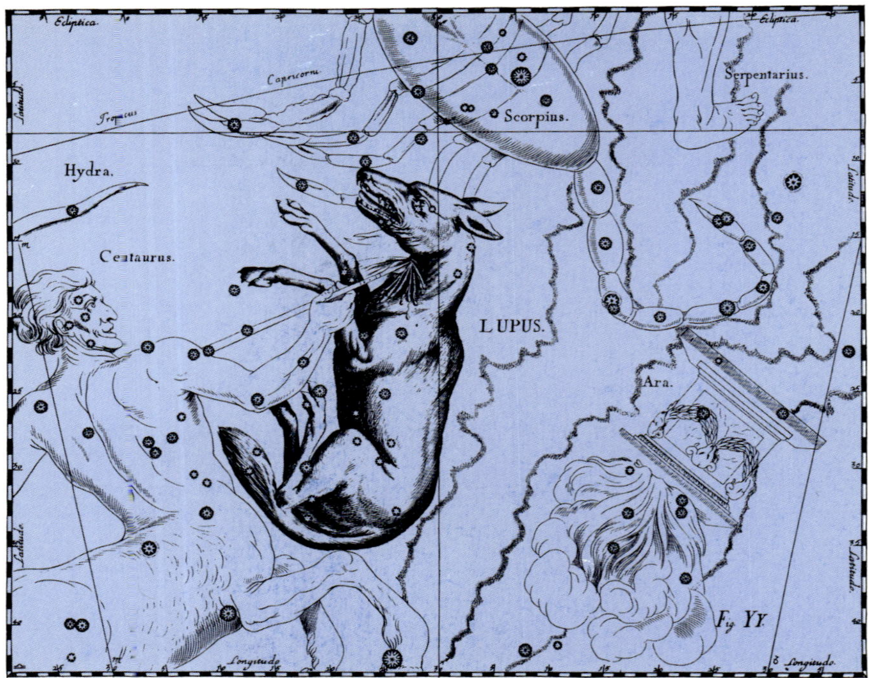

July

Globulars Galore

THERE are four outstanding globular clusters in the sky, and the Southern Hemisphere is lucky enough to possess three of them. In fact all four can be seen from southern latitudes during July, when M13 in Hercules and M22 in Sagittarius culminate at about the same time and Omega Centauri and 47 Tucanae straddle the south celestial pole.

These rich balls of stars are a big cut above ordinary open star clusters. Globulars typically contain hundreds of thousands of the oldest stars known; open clusters usually number their stars in the mere dozens or hundreds, and they are mostly young. Globulars typically have 10 times the diameter of open clusters — but unfortunately they average roughly 10 times the distance, too.

Omega (ω) Centauri is unquestionably the finest globular cluster in the sky. You won't have a problem finding it, since extending a line from Beta (β) through Epsilon (ε) Centauri takes you right to it. Mind you, it should only be under the worst conditions of city lights or bright moonlight that this trick would be necessary, as Omega is easily visible to the naked eye.

More than 1,800 years ago Ptolemy saw Omega as a hazy 4th-magnitude star, and he included it as such in his famous star catalog. The designation Omega was given to the star by Johann Bayer in his star atlas of 1603, *Uranometria;* it was he who first assigned Greek letters to stars within a constellation. However, it was not until Edmond Halley's observations in 1677 that its true nature as a star cluster was realized.

There is no word other than *superb* to describe the view of Omega Centauri through a telescope on a dark night. Under moderate magnification it fills the field of view with countless points of light. To me it looks like a multitude of fireflies streaming away in every direction. When I first started out in astronomy I spent hours gazing at this cluster wondering how all the stars could exist so closely together. "Weren't they always running into each other?" I thought. Only many years later did I learn that even at its dense center, the average distance between stars is still about 1,500 times the distance of Pluto from the Sun. So there is still plenty of room!

But inhabitants of a planet within Omega Centauri would find the heavens filled with incomparable splendor. There would be no real night, and the brilliance of their skies would probably make them unaware of the vastness of the universe beyond.

The far southern sky has another of these globular wonders — 47 Tucanae, on the western edge of the Small Magellanic Cloud. Although it's a little fainter than Omega, 47 Tuc stands out more to the naked eye because, unlike Omega, it lies well away from the plane of the Milky Way in a region of few foreground stars. However, both Omega and 47 Tuc are at about the same distance (around 20,000 light-years), so comparing how big they look in the sky gives a good idea of their true relative sizes.

NGC 5139 (Omega Centauri)

NGC 104 (47 Tucanae)

Before I first saw these clusters I tended to think that all globulars were alike. But comparing Omega Centauri and 47 Tucanae reveals their distinct architectures. Whereas the former has a smooth gradation of stars from center to edge, the latter's bright, condensed, central core almost hits you in the eye. The halo of 47 Tuc is also very well defined, featuring one especially bright star and streamers of fainter stars shooting off in several directions.

For another completely different perspective on globulars, though, take a look at M22, about 2° northeast of Lambda (λ) Sagittarii. This cluster is easily visible in binoculars as a smoky patch of light and is easy to resolve in a telescope. Yet because its central region is relatively uniform it lacks the grandeur of its brighter cousins farther south. M22 appears distinctly elliptical, whereas Omega Centauri and 47 Tuc are quite round. At a distance of about 10,000 light-years, M22 is significantly closer than its southern counterparts, and among the nearest of all globulars.

The finest globular cluster north of the equator is M13, just visible to the naked eye about a third of the way from Eta (η) to Zeta (ζ) Herculis. From midsouthern latitudes it never rises high enough for skywatchers to appreciate its full beauty. In this case, at least, the lands that harbor the most astronomers are afforded a decent view.

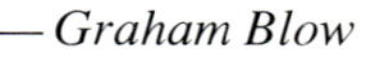
— *Graham Blow*

M22 in Sagittarius

Interesting Objects

Object	Constellation	Type	Right Ascension	Dec.	Size/Sep./Period	Magnitude
Coalsack Nebula	Crux	DkN	12^h 53^m	–63° 03′	7°×5°	
M4	Scorpius	GC	16^h 23^m.6	–26° 32′	26′	5.9
M5	Serpens	GC	15^h 18^m.6	+02° 05′	17′	5.8
M6 (Butterfly Cluster)	Scorpius	OC	17^h 40^m	–32° 13′	15′	4.2
M7	Scorpius	OC	17^h 53^m.9	–34° 49′	80′	3.3
M8 (Lagoon Nebula)	Sagittarius	DN	18^h 03^m.8	–24° 23′	90′×40′	6
M10	Ophiuchus	GC	16^h 57^m.1	–04° 06′	15′	6.6
M12	Ophiuchus	GC	16^h 47^m.2	–01° 57′	14′	6.6
M19	Ophiuchus	GC	17^h 02^m.6	–26° 16′	14′	7.2
M20 (Trifid Nebula)	Sagittarius	DN	18^h 02^m.6	–23° 02′	29′×27′	8
M22	Sagittarius	GC	18^h 36^m.4	–23° 54′	24′	5.1
M80	Scorpius	GC	16^h 17^m.0	–22° 59′	9′	7.2
NGC 6067	Norma	OC	16^h 13^m.2	–54° 13′	13′	5.6
NGC 6210	Hercules	PN	16^h 44^m.5	+23° 49′	0′.2	9
ω Centauri (NGC 5139)	Centaurus	GC	13^h 26^m.8	–47° 29′	36′	3.7
47 Tucanae (NGC 104)	Tucana	GC	00^h 24^m.1	–72° 05′	31′	4

DN = diffuse nebula; DS = double or multiple star; DkN = dark nebula; G = galaxy; GC = globular cluster; OC = open cluster; PN = planetary nebula; VS = variable star

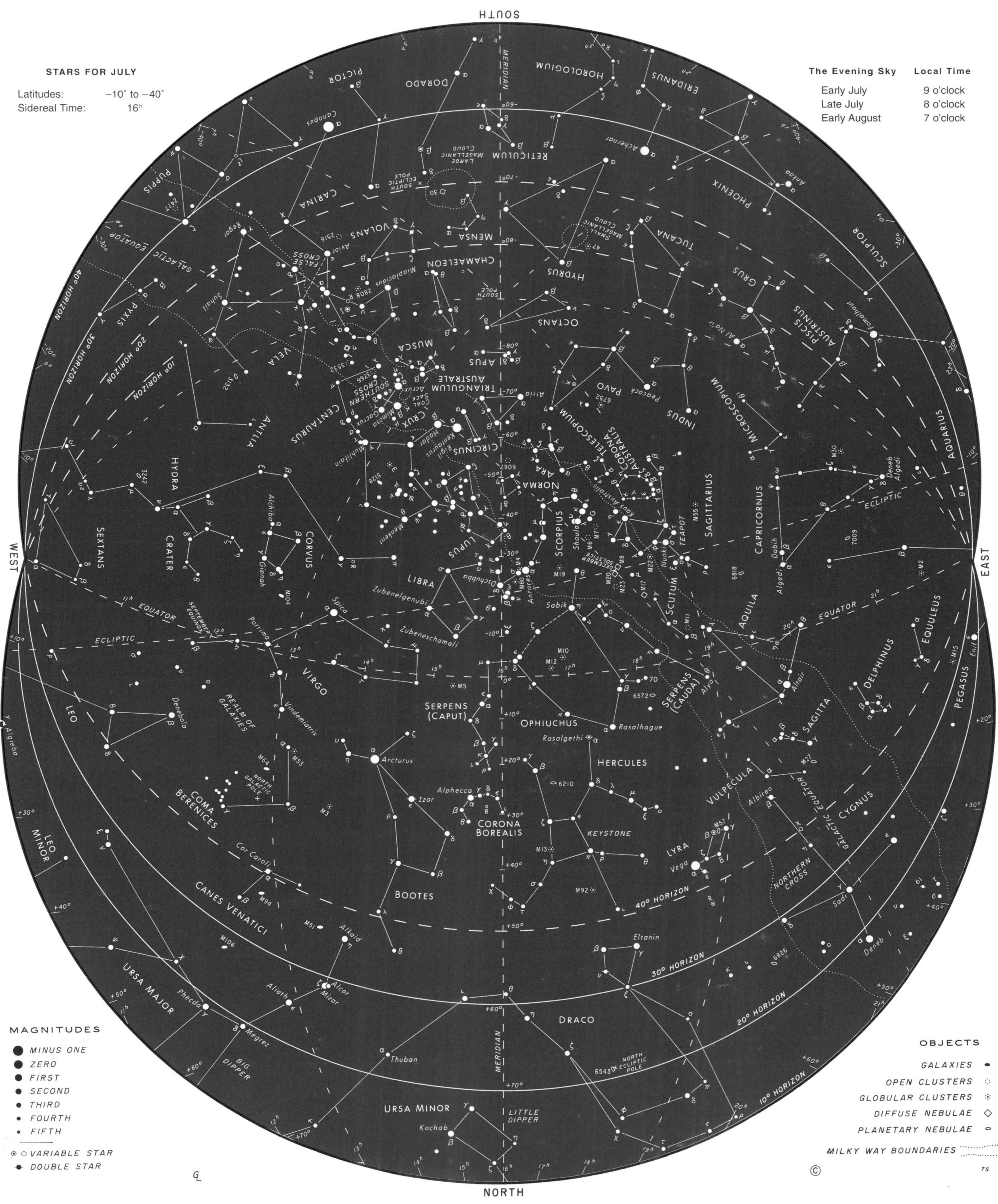
STARS FOR JULY
Latitudes: −10° to −40°
Sidereal Time: 16h
The Evening Sky Local Time
Early July 9 o'clock
Late July 8 o'clock
Early August 7 o'clock
SOUTH
NORTH
EAST
WEST
MAGNITUDES
MINUS ONE
ZERO
FIRST
SECOND
THIRD
FOURTH
FIFTH
VARIABLE STAR
DOUBLE STAR
OBJECTS
GALAXIES
OPEN CLUSTERS
GLOBULAR CLUSTERS
DIFFUSE NEBULAE
PLANETARY NEBULAE
MILKY WAY BOUNDARIES
SCORPIUS
SAGITTARIUS
OPHIUCHUS
HERCULES
LIBRA
VIRGO
BOOTES
CORONA BOREALIS
SERPENS (CAPUT)
SERPENS (CAUDA)
LYRA
CYGNUS
AQUILA
DRACO
URSA MINOR
URSA MAJOR
CANES VENATICI
COMA BERENICES
LEO
CORVUS
CRATER
HYDRA
CENTAURUS
CRUX
CARINA
VELA
ECLIPTIC
EQUATOR
MERIDIAN

AUGUST

Astronomers have argued for years about what might be at the heart of our galaxy. Radio astronomers have long known of a powerful source of radio emission close to the center. They call it Sagittarius A* (pronounced "A star"), and conjecture that it might be the result of matter falling into a massive black hole. But infrared astronomers see things quite differently. For them the cluster of powerful infrared-emitting objects detected close to the Milky Way's core suggests that our galaxy's powerhouse is a dense star cluster no more than 10,000 times the mass of our Sun.

Although we cannot see this cluster, we can estimate its position using this month's sky chart. From Gamma (γ) Sagittarii draw a line to the star cluster M6. The apex of the almost equilateral triangle formed with the dashed line labeled Galactic Equator is 0° galactic longitude.

M8 in Sagittarius

M20 in Sagittarius

Something for Everyone in Sagittarius

AUGUST at midsouthern latitudes is a time of long, cold winter nights and unparalleled views of the center of our galaxy almost directly overhead. Away from city lights the Milky Way's stretch from Sagittarius down to the southwest is breathtaking, and for the early Polynesian navigators it undoubtedly provided an important reference in their travels.

Polynesian lore sometimes refers to the Milky Way as Mongoroa, or "long shark." It is easy to see the connection, with island lifestyles based so much on the sea. Another myth speaks of the Milky Way as Ikaroa, one of the offspring of the Sky Father and Earth Mother. The stars in turn were the children of Ikaroa, perhaps foreshadowing our knowledge of the true nature of the Milky Way. It was Ikaroa's duty to control the younger stars, but occasionally some of them would stray. Then their elders would lash out, causing them to become *matakokiri* (meteors).

The ancient Polynesians could not have foretold how the constellation Sagittarius would today be such a fruitful playground for both professional and amateur astronomers. For the professionals, its fascination lies in its marking the very center of our galaxy — the most violent and energetic part. But seeing into this maelstrom is easier said than done, because the vast, dense dust clouds spanning western Sagittarius block our galaxy's core from view, in visible light. Radio and infrared observations are the most effective means of slicing through this celestial murk.

Interesting Objects

Object	Constellation	Type	Right Ascension	Dec.	Size/Sep./ Period	Magnitude
M6 (Butterfly Cluster)	Scorpius	OC	$17^h\ 40^m$	–32° 13′	15′	4.2
M7	Scorpius	OC	$17^h\ 53^m.9$	–34° 49′	80′	3.3
M8 (Lagoon Nebula)	Sagittarius	DN	$18^h\ 03^m.8$	–24° 23′	90′ × 40′	6
M10	Ophiuchus	GC	$16^h\ 57^m.1$	–04° 06′	15′	6.6
M11 (Wild Duck Cluster)	Scutum	OC	$18^h\ 51^m.1$	–06° 16′	14′	5.8
M12	Ophiuchus	GC	$16^h\ 47^m.2$	–01° 57′	14′	6.6
M17 (Omega Nebula)	Sagittarius	DN	$18^h\ 20^m.8$	–16° 11′	46′ × 37′	7
M19	Ophiuchus	GC	$17^h\ 02^m.6$	–26° 16′	14′	7.2
M20 (Trifid Nebula)	Sagittarius	DN	$18^h\ 02^m.6$	–23° 02′	29′ × 27′	8
M22	Sagittarius	GC	$18^h\ 36^m.4$	–23° 54′	24′	5.1
M23	Sagittarius	OC	$17^h\ 56^m.8$	–19° 01′	27′	5.5
M55	Sagittarius	GC	$19^h\ 40^m.0$	–30° 58′	19′	7.0
M57 (Ring Nebula)	Lyra	PN	$18^h\ 53^m.6$	+33° 02′	1′	9
NGC 6210	Hercules	PN	$16^h\ 44^m.5$	+23° 49′	0′.2	9
NGC 6572	Ophiuchus	PN	$18^h\ 12^m.1$	+06° 51′	0′.1	9
NGC 6752	Pavo	GC	$19^h\ 10^m.9$	–59° 59′	20′.4	5.4

DN = diffuse nebula; DS = double or multiple star; DkN = dark nebula; G = galaxy; GC = globular cluster; OC = open cluster; PN = planetary nebula; VS = variable star

For amateur astronomers, Sagittarius — easily recognizable by its familiar "Teapot" asterism — is a treasure-trove of celestial wonders, containing 49 globular and galactic star clusters as well as many nebulae. As a teenager I found it a thrill to sweep my father's binoculars along the Milky Way in Sagittarius, examining each knot and swirl, which just begged to be investigated further. My first telescope, a simple 60-mm refractor, gave me the means to examine these curiosities in more detail. I could see M8, the famous Lagoon Nebula, fairly clearly, appreciating the cometlike nebulosity that swirls in bright and dark bands around and eastwards from the 7th-magnitude star 9 Sagittarii. Just across the ecliptic was another well-known nebula from Charles Messier's 18th-century catalog of objects — the Trifid, or M20 — only this one was rather more difficult to make out. I had hoped to discern its three tortuous dark lanes, which in photographs, such as the one at upper right, are so strikingly superimposed on the bright, gaseous background. But these were beyond the reach of my small-aperture instrument.

Perhaps more easily seen than the nebulae are the globular and galactic star clusters, of which the open cluster M23 (close to the border with Ophiuchus) is of special note. Its stars swirl in loops, chains, and circlets — and a 6th-magnitude star to the northwest shines like a diamond in the cluster's ring.

If you like your viewing in concentrated doses, Sagittarius provides generous examples of just about everything our galaxy has produced. The Archer, as this constellation represents, himself makes a great target for astronomers of all levels.

— Graham Blow

M55 in Sagittarius

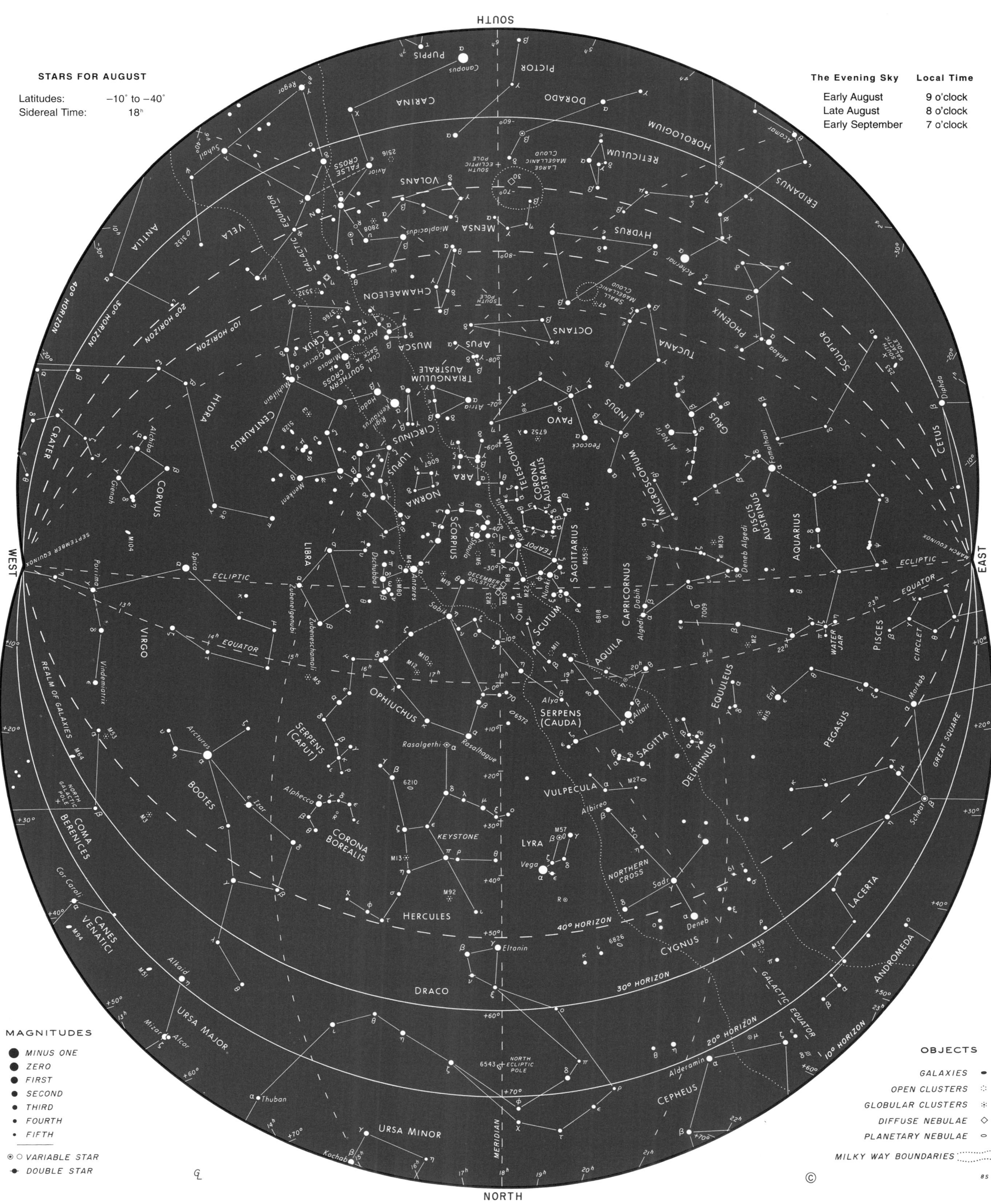
STARS FOR AUGUST
Latitudes: −10° to −40°
Sidereal Time: 18h
The Evening Sky Local Time
Early August 9 o'clock
Late August 8 o'clock
Early September 7 o'clock
SOUTH
NORTH
EAST
WEST
MAGNITUDES
MINUS ONE
ZERO
FIRST
SECOND
THIRD
FOURTH
FIFTH
VARIABLE STAR
DOUBLE STAR
OBJECTS
GALAXIES
OPEN CLUSTERS
GLOBULAR CLUSTERS
DIFFUSE NEBULAE
PLANETARY NEBULAE
MILKY WAY BOUNDARIES

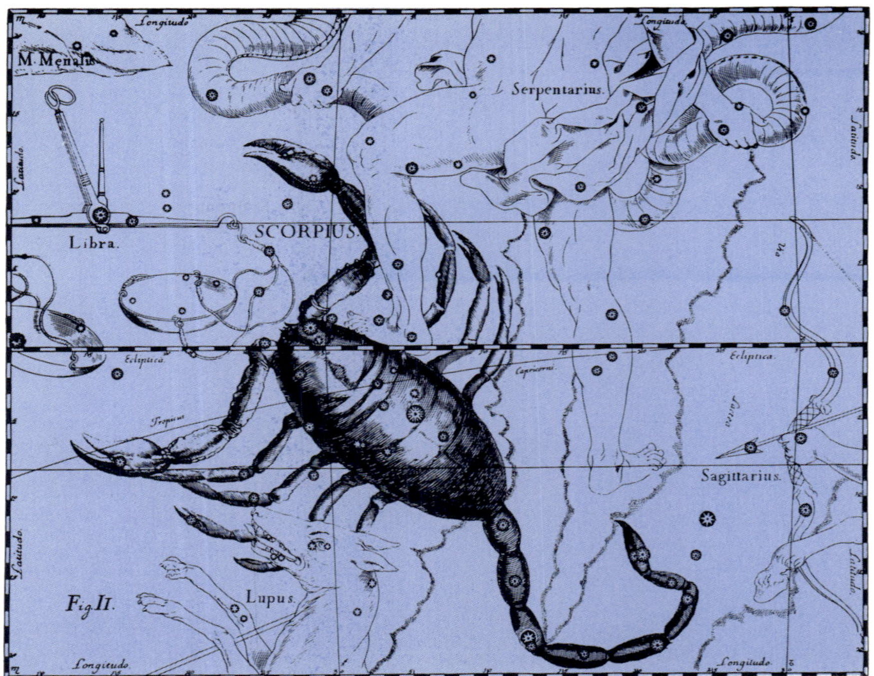

SEPTEMBER

The constellation of Scorpius is a gold mine of interesting objects that can keep an observer happy for hours. Just north of the Scorpion's stinger are the two galactic clusters Messiers 6 and 7, included in Ptolemy's *Almagest* because of their easy visibility to the naked eye. M7 is a true binocular object, being over 1° across. On its southwestern side a stream of stars seems to spiral out from its center. M6, on the other hand, is much smaller and looks rather like an oval with one end pushed askew. Like M7, its stars are mostly blue-white, except for one that seems distinctly orange-red.

Forming a triangle with M6 and M7 are Lambda (λ) and Upsilon (υ) Scorpii, at the end of the Scorpion's tail. In Arabic these were known as Al Shaulah and Al Las'ah, respectively, both names meaning "the Sting." These two hot, blue stars belong to the huge Scorpius-Centaurus Association of young stars that provides many of this sector's leading lights. Lambda and Upsilon have the same absolute magnitude (luminosity); thus, it is their differing distances of 300 and 550 light-years, respectively, that account for their different apparent brightnesses.

M6 in Scorpius

M7, near M6

Following the curve of bright stars around from the stinger brings us to Theta (θ) and Eta (η) Scorpii. Almost in a line with them are Zeta^1 (ζ^1) and Zeta^2 (ζ^2) Scorpii, whose color difference is dazzling. The hot, blue Zeta^1 and cool, orange Zeta^2 contrast nicely with the fainter yellow of SAO 227392 (not plotted), with which they form an almost equilateral triangle.

A half degree to the north of the Zeta pair is the beautiful open cluster NGC 6231, a gorgeous collection of white and yellow stars that seem to sparkle in flowing lines and curves. At magnitude 2.6, it's distinct with the unaided eye, and under low magnification it reminds me of a miniature Pleiades, its precious stones shimmering against a background of black velvet.

About 5° due north of the two Zetas is the stunning naked-eye double star Mu (μ) Scorpii. Its components are almost the same brightness and remind me of two siblings huddling together. Indeed, they appear in the sky lore of the Cook Islands as Piri-ere-ua, the Inseparable Ones — two children fleeing their wicked parents.

The rest of Scorpius is just as interesting as its tail. In and around its head, bright orange Antares, the globular cluster M4, and the fine telescopic double stars Beta (β) and Nu (ν) Scorpii add to the Scorpion's reputation as one of the skygazer's best friends.

— *Graham Blow*

THE SCORPION'S STING

FOR the beginner new to the sky only a few constellations have shapes easy to recognize right away. The Big Dipper, Orion, the Southern Cross, and Scorpius are perhaps the four best known.

Because Scorpius lies well south of the celestial equator it is best seen from midsouthern latitudes, where its distinctive curve of stars passes overhead. In the mythology of the Pacific Islands, these stars are known as Te-matua-o-Maui, or the Fishhook of Maui.

In Polynesian lore Maui was a demigod who fished with a hook made from the jawbone of a whale. On one occasion Maui cast his hook far, and it snagged on the bottom of the sea. Pulling with all his might, Maui and his fishhook brought up the islands now known as New Zealand. But he pulled so hard that his line snapped, and both line and hook catapulted backwards into the sky, where we see them today.

While unknown in the Pacific, scorpions were common in ancient Europe and the Mideast. So it is not surprising that people there saw the constellation's distinctive curve of stars as evoking a scorpion's tail and stinger.

One tale has it that Scorpius was sent by Gaia, the Earth Goddess, to punish the mighty hunter Orion, who had arrogantly boasted that he would kill all the animals on Earth. This so incensed Gaia that she instructed the Scorpion to sting Orion on the heel. The giant then sank, mortally wounded, to the Earth. We see this tableau reenacted in the sky year after year, for as Orion sets in the west, the Scorpion rises in the east.

INTERESTING OBJECTS

Object	Constellation	Type	Right Ascension	Dec.	Size/Sep./ Period	Magnitude
Small Magellanic Cloud	Tucana	G	00^h 52^m.7	−72° 50′	5°×3°	2.8
M2	Aquarius	GC	21^h 33^m.5	−00° 49′	13′	6.5
M6 (Butterfly Cluster)	Scorpius	OC	17^h 40^m	−32° 13′	15′	4.2
M7	Scorpius	OC	17^h 53^m.9	−34° 49′	80′	3.3
M8 (Lagoon Nebula)	Sagittarius	DN	18^h 03^m.8	−24° 23′	90′ × 40′	6
M11 (Wild Duck Cluster)	Scutum	OC	18^h 51^m.1	−06° 16′	14′	5.8
M15	Pegasus	GC	21^h 30^m.0	+12° 10′	12′	6.4
M17 (Omega Nebula)	Sagittarius	DN	18^h 20^m.8	−16° 11′	46′ × 37′	7
M20 (Trifid Nebula)	Sagittarius	DN	18^h 02^m.6	−23° 02′	29′ × 27′	8
M30	Capricornus	GC	21^h 40^m.4	−23° 11′	11′	7.5
M55	Sagittarius	GC	19^h 40^m.0	−30° 58′	19′	7.0
NGC 6752	Pavo	GC	19^h 10^m.9	−59° 59′	20′.4	5.4
NGC 7009 (Saturn Nebula)	Aquarius	PN	21^h 04^m.2	−11° 22′	2′	8
δ Apodis	Apus	DS	16^h 20^m.3	−78° 42′	103″	4.7, 5.1
α Capricorni (Algedi)	Capricorn	DS	20^h 18^m.1	−12° 33′	378″	3.6, 4.2
μ Scorpii	Scorpius	VS/DS	16^h 51^m.9	−38° 05′	348″	3.0–3.3, 3.6
47 Tucanae (NGC 104)	Tucana	GC	00^h 24^m.1	−72° 05′	31′	4

DN = diffuse nebula; DS = double or multiple star; DkN = dark nebula; G = galaxy; GC = globular cluster; OC = open cluster; PN = planetary nebula; VS = variable star

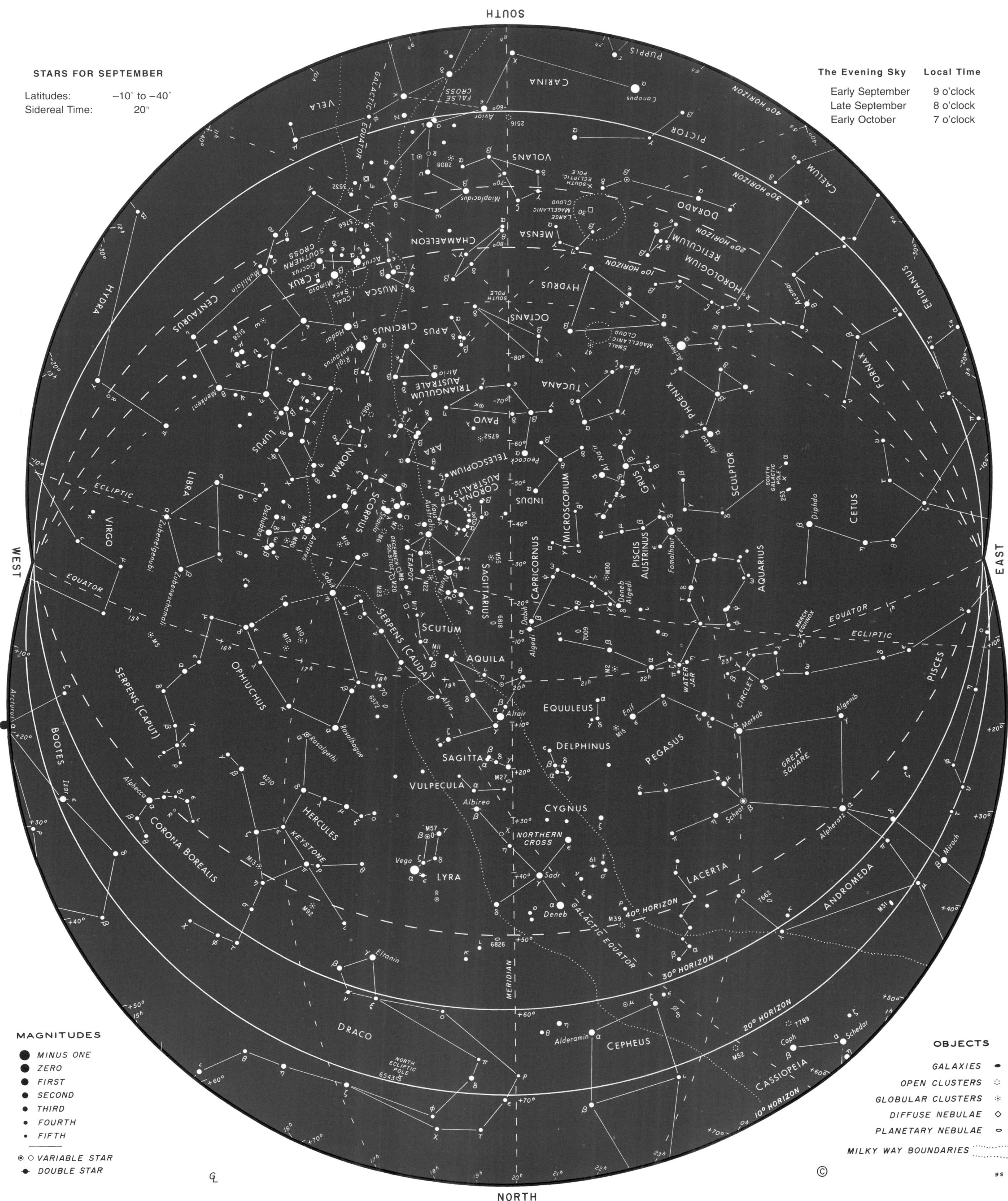
STARS FOR SEPTEMBER
Latitudes: −10° to −40°
Sidereal Time: 20h
The Evening Sky | Local Time
Early September | 9 o'clock
Late September | 8 o'clock
Early October | 7 o'clock
SOUTH
NORTH
WEST
EAST
MAGNITUDES
MINUS ONE
ZERO
FIRST
SECOND
THIRD
FOURTH
FIFTH
VARIABLE STAR
DOUBLE STAR
OBJECTS
GALAXIES
OPEN CLUSTERS
GLOBULAR CLUSTERS
DIFFUSE NEBULAE
PLANETARY NEBULAE
MILKY WAY BOUNDARIES
CARINA
PUPPIS
VELA
PICTOR
VOLANS
CAELUM
DORADO
MENSA
CHAMAELEON
RETICULUM
HOROLOGIUM
CRUX
SOUTHERN CROSS
FALSE CROSS
MUSCA
HYDRUS
OCTANS
ERIDANUS
HYDRA
CENTAURUS
CIRCINUS
APUS
TRIANGULUM AUSTRALE
TUCANA
PHOENIX
FORNAX
LUPUS
NORMA
ARA
PAVO
TELESCOPIUM
CORONA AUSTRALIS
INDUS
MICROSCOPIUM
GRUS
SCULPTOR
CETUS
LIBRA
VIRGO
SCORPIUS
SAGITTARIUS
CAPRICORNUS
PISCIS AUSTRINUS
AQUARIUS
SCUTUM
SERPENS (CAUDA)
AQUILA
OPHIUCHUS
SERPENS (CAPUT)
EQUULEUS
DELPHINUS
PEGASUS
GREAT SQUARE
CIRCLET
WATER JAR
PISCES
SAGITTA
VULPECULA
CYGNUS
NORTHERN CROSS
LYRA
HERCULES
KEYSTONE
CORONA BOREALIS
BOOTES
LACERTA
ANDROMEDA
DRACO
CEPHEUS
CASSIOPEIA
TEAPOT
ECLIPTIC
EQUATOR
GALACTIC EQUATOR
MERIDIAN
SOUTH ECLIPTIC POLE
NORTH ECLIPTIC POLE
SOUTH GALACTIC POLE
LARGE MAGELLANIC CLOUD
SMALL MAGELLANIC CLOUD
MARCH EQUINOX
DECEMBER SOLSTICE
10° HORIZON
20° HORIZON
30° HORIZON
40° HORIZON
Canopus
Avior
Miaplacidus
Achernar
Acamar
Hadar
Rigil Kentaurus
Mimosa
Gacrux
Acrux
Atria
Peacock
Al Na'ir
Fomalhaut
Diphda
Ankaa
Antares
Shaula
Kaus Australis
Dschubba
Zubenelgenubi
Zubeneschamali
Sabik
Nunki
Dabih
Algedi
Deneb Algedi
Enif
Markab
Scheat
Algenib
Alpheratz
Mirach
Altair
Alya
Albireo
Vega
Sadr
Deneb
Rasalhague
Rasalgethi
Alphecca
Izar
Arcturus
Menkent
Muhlifain
Eltanin
Alderamin
Caph
Schedar

OCTOBER

A Traverse of the Southern Ocean

THERE is something decidedly fishy about the October Southern Hemisphere sky. With Piscis Austrinus high near the meridian in the early evening — surrounded by Aquarius, the Water Carrier; Capricornus, the Sea Goat; Grus, the Crane; and Cetus, the Whale — the heavens present a distinctly wet motif that extends all the way down to the River Eridanus in the east and Pisces, the Fishes, in the northeast.

This celestial ocean and its inhabitants resulted from the close association of early civilizations with the sea. Many cities were built on natural harbors and depended on a thriving sea trade with distant lands. Undoubtedly for them (as it still is for us!), the sea was a vast and mysterious place populated by strange creatures. Understandably, early voyagers recorded some of what they saw, or imagined they saw, in the constellations.

Take Cetus, for example, spanning the equator in the northeast. Although today we think of it as a whale, this constellation's early association was with any sort of large and fearsome sea creature. Nowadays the creature swims in a rather barren area of sky and possesses no star brighter than 2nd magnitude. It has neither open nor globular clusters and contains only one planetary nebula, 8th-magnitude NGC 246.

If Cetus does have a claim to fame (aside from its many distant galaxies), it must be for the two red variable-star prototypes it houses: UV Ceti, the first flare star discovered (not plotted on the chart), and Omicron Ceti — better known as Mira — the archetypal long-period variable.

Immediately north of Cetus lies Pisces, while to its west and south swims Piscis Austrinus (also known as Piscis Australis; the two forms have been used interchangeably). Although the northern Pisces is a relatively large constellation, its southern cousin is really much more of a sprat. But at least the Southern Fish does contain a first-magnitude star, Fomalhaut. Named from the Arabic *Fum-al-Hut* (mouth of the fish), it is depicted on old star charts as the end of the stream poured by Aquarius. It is also sometimes called the "Solitary One," because it stands, lighthouselike, in an otherwise completely undistinguished region.

Immediately to its south, and perhaps hungrily eyeing its fishy neighbor, perches Grus, the Crane. The distinctive line of naked-eye stars stretching from Gamma (γ) through Beta (β) to Zeta (ζ) Gruis forms an easily recognized Y shape with 2nd-magnitude Alpha (α) off to the west. The ancient Arabs included the stars of Grus amongst those of the Southern Fish; the constellation we know today was introduced by Johann Bayer in 1603. Being well away from the Milky Way, Grus is devoid of star clusters or nebulae, and even its galaxies are faint and uninspiring.

Interesting Objects

Object	Constellation	Type	Right Ascension	Dec.	Size/Sep./ Period	Magnitude
Large Magellanic Cloud	Dorado	G	05^h 23^m.6	–69° 45′	11°×9°	0.6
Small Magellanic Cloud	Tucana	G	00^h 52^m.7	–72° 50′	5°×3°	2.8
M2	Aquarius	GC	21^h 33^m.5	–00° 49′	13′	6.5
M15	Pegasus	GC	21^h 30^m.0	+12° 10′	12′	6.4
M30	Capricornus	GC	21^h 40^m.4	–23° 11′	11′	7.5
M55	Sagittarius	GC	19^h 40^m.0	–30° 58′	19′	7.0
NGC 253 (Sculptor Galaxy)	Sculptor	G	12^h 43^m.7	+11° 33′	7′×6′	8.8
NGC 6752	Pavo	GC	19^h 10^m.9	–59° 59′	20′.4	5.4
NGC 7009 (Saturn Nebula)	Aquarius	PN	21^h 04^m.2	–11° 22′	2′	8
δ Apodis	Apus	DS	16^h 20^m.3	–78° 42′	103″	4.7, 5.1
α Capricorni (Algedi)	Capricorn	DS	20^h 18^m.1	–12° 33′	378″	3.6, 4.2
30 Doradus (Tarantula Nebula, NGC 2070)	Dorado	DN	05^h 38^m.7	–69° 06′	40′×25′	5
β Tucanae	Tucana	DS	00^h 31^m.5	–62° 58′	27″.1	4.4, 4.8
47 Tucanae (NGC 104)	Tucana	GC	00^h 24^m.1	–72° 05′	30′.9	4
κ Pavonis	Pavo	VS	18^h 56^m.9	–67° 14′	9.1 days	3.94–4.75

DN = diffuse nebula; DS = double or multiple star; DkN = dark nebula; G = galaxy; GC = globular cluster; OC = open cluster; PN = planetary nebula; VS = variable star

NGC 253 in Sculptor

M30 in Capricornus

However, right next door on the shores of our celestial ocean is Sculptor, host to one of the best galaxies in the southern skies after the Magellanic Clouds. NGC 253 is marked on our chart right beside the south galactic pole. An excellent example of an almost edgewise spiral, the "Sculptor Galaxy," as it is known, is readily visible in binoculars as a chalky streak about ½° long, even in moonlight. It grades in brightness uniformly from center to edge.

Diving from Sculptor across Piscis Austrinus and Aquarius brings us to Capricornus. The Sea Goat (or, sometimes, Goat-Fish) originated in Sumeria but became associated with the Greek myth of Pan. While quietly playing his pipes beside a river one day, Pan (a goat) was terrified by the sudden appearance of the demon Typhon. In a panic (this word comes from Pan's own name), Pan dived into the river to change himself into a fish and swim away. However, when we panic we rarely do anything right, and, with everything happening so quickly, only Pan's hindquarters completed the transformation — his forequarters remained those of a goat. He was stuck that way ever after. Just one of the strange inhabitants — and delightful surprises — in the celestial sea.

— *Graham Blow*

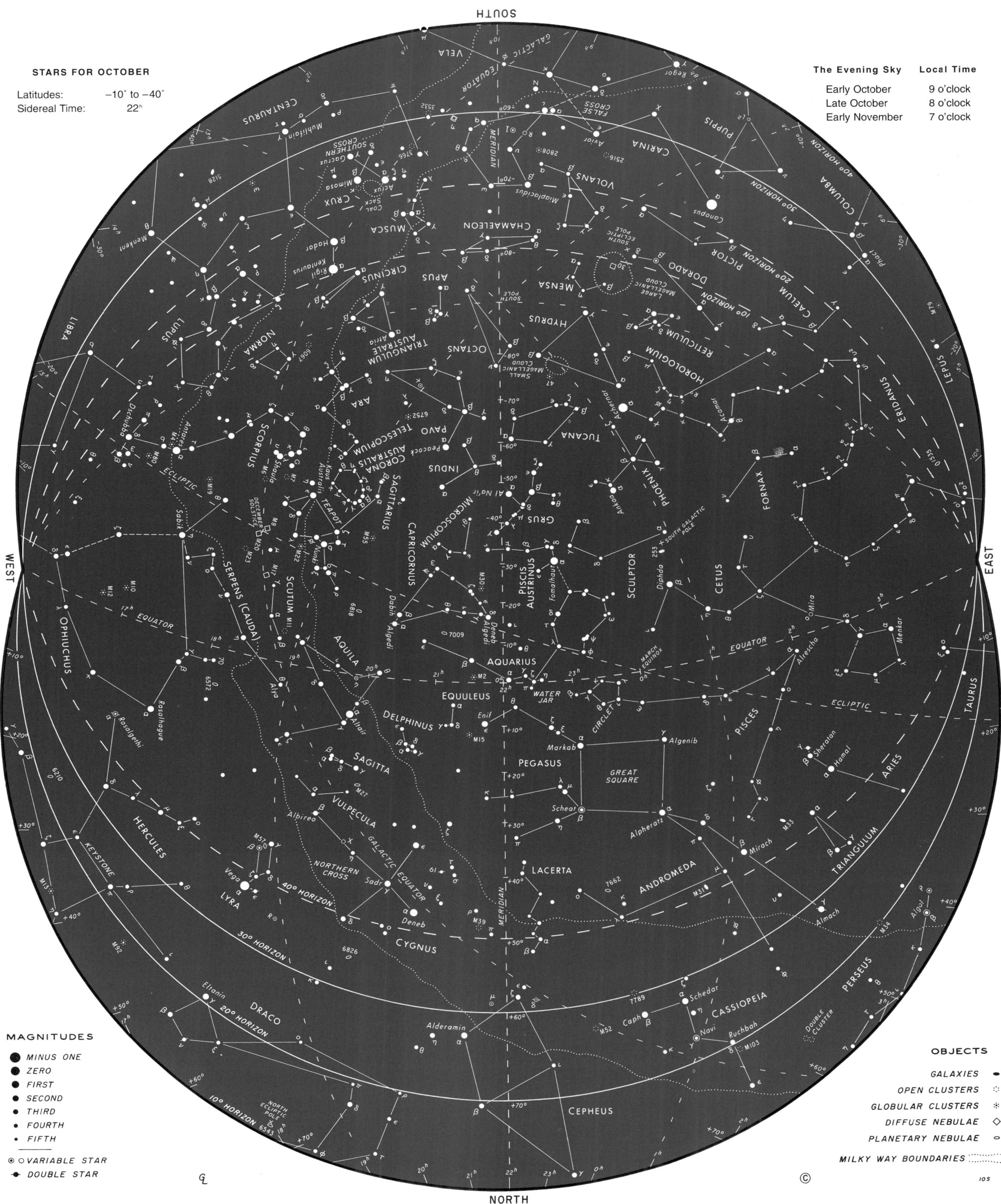
STARS FOR OCTOBER
Latitudes: -10° to -40°
Sidereal Time: 22h
The Evening Sky / Local Time
Early October 9 o'clock
Late October 8 o'clock
Early November 7 o'clock
SOUTH
NORTH
EAST
WEST
MAGNITUDES
MINUS ONE
ZERO
FIRST
SECOND
THIRD
FOURTH
FIFTH
VARIABLE STAR
DOUBLE STAR
OBJECTS
GALAXIES
OPEN CLUSTERS
GLOBULAR CLUSTERS
DIFFUSE NEBULAE
PLANETARY NEBULAE
MILKY WAY BOUNDARIES
VELA
CENTAURUS
SOUTHERN CROSS
CRUX
CARINA
PUPPIS
COLUMBA
VOLANS
MUSCA
CHAMAELEON
CIRCINUS
APUS
MENSA
DORADO
PICTOR
CAELUM
LIBRA
LUPUS
NORMA
TRIANGULUM AUSTRALE
OCTANS
HYDRUS
RETICULUM
HOROLOGIUM
LEPUS
ERIDANUS
ARA
PAVO
TUCANA
SCORPIUS
TELESCOPIUM
CORONA AUSTRALIS
INDUS
PHOENIX
FORNAX
SAGITTARIUS
MICROSCOPIUM
GRUS
CAPRICORNUS
PISCIS AUSTRINUS
SCULPTOR
CETUS
SERPENS (CAUDA)
SCUTUM
OPHIUCHUS
AQUILA
AQUARIUS
EQUULEUS
DELPHINUS
PISCES
TAURUS
SAGITTA
PEGASUS
GREAT SQUARE
ARIES
HERCULES
VULPECULA
TRIANGULUM
LYRA
NORTHERN CROSS
LACERTA
ANDROMEDA
CYGNUS
PERSEUS
DRACO
CASSIOPEIA
CEPHEUS
EQUATOR
ECLIPTIC
GALACTIC EQUATOR
MERIDIAN
10° HORIZON
20° HORIZON
30° HORIZON
40° HORIZON

NOVEMBER

THE SOUTHERN SKY'S ATTIC

THE sky along the southern Milky Way, which looms high overhead on winter evenings, contains a bounty of brilliant stars and an incredible variety of deep-sky objects, all of which once inspired me to call it a "celestial candy store."

However, as we head away from that luminous band toward the south celestial pole, the sky grows considerably more barren. The area around the pole itself has been called "the South Polar Pit" by some observers. It even lacks a prominent pole star like the North's Polaris; the nearest naked-eye star is Sigma (σ) Octantis, which at magnitude 5.47 is *barely* visible to the naked eye. Moreover, at declination –89°, the star is one whole degree from the pole, making things more difficult for Southern Hemisphere observers trying to align their telescopes accurately. (Polaris is currently only ¾° from the north celestial pole.)

The zone of sky some 40° in radius surrounding the southern pole remained unknown to the ancients, so it has no legacy of classical constellations steeped in mythology. Indeed, it wasn't until some five centuries ago, with the coming of the Age of Exploration, that Europeans began sailing south of the equator regularly, particularly around the southern tip of Africa. These navigators, mainly Portuguese, were essentially the first Western people to see those "new" southern stars.

Another hundred years passed before star watchers colonized this large blank area with new post-classical constellations. The first dozen were formed by the Dutch explorer Pieter Dirckszoon Keyser in 1596 on a voyage to the East Indies, where his country already had significant interests. Following tradition, most of his star patterns represented living creatures, this time with a partiality to birds. His feathered contingent includes Apus, the Bird of Paradise; Grus, the Crane; Pavo, the Peacock; Phoenix, the mythical Phoenix Bird; and Tucana, the Toucan. The remaining Keyser groups are Chamaeleon, the Chameleon; Dorado, the Gold Fish; Hydrus, the male Water Snake; Indus, the New World Indian; Musca, the Fly; and Triangulum Australe, the Southern Triangle. The last is the only inanimate Keyser constellation.

Notice that this bunch is not grouped symmetrically around the south celestial pole, but is offset toward the March equinox. Why? Precession. During the so-called classical period about 2,500 to 2,000 years ago, the pole lay among the stars of present-day Hydrus — in fact, in the vicinity of the Small Magellanic Cloud. Therefore the area of sky invisible or poorly seen from Mideast-Mediterranean locales was centered on Hydrus, at the middle of the Keyser collection.

Although this group of 12 appeared on at least three Dutch celestial globes around the beginning of the 17th century, it was their inclusion in Johann Bayer's epochal *Uranometria* atlas of 1603 that gave these new constellations "official" status. This has given rise to the often repeated assertion that it was Bayer who introduced them. It depends on one's definition of "introduction" versus "recognition." Many constellations were created over the centuries — some patently ridiculous — and never recognized by anyone other than their creators.

Nicolas Louis de Lacaille

NGC 7009 in Aquarius

The Keyser patterns used up most of the brighter naked-eye stars around the south celestial pole, but this did not deter the French astronomer Nicolas Louis de Lacaille (1713–1762) from filling the remaining gaps with 14 faint new constellations in the early 1750s during a sojourn at the Cape of Good Hope. All of his patterns represented inanimate objects, largely scientific devices. Among others, Lacaille added Antlia, the Air Pump; Caelum, the Engraving Tool; Fornax, the Chemical Furnace; Mensa, South Africa's Table Mountain; Microscopium, the Microscope; Octans, the Octant; Pictor, the Painter's Easel; Sculptor, the Sculptor's Apparatus; Telescopium, the Telescope; and Reticulum, the eyepiece reticle in Lacaille's own small telescope that served as an aid in measuring star positions.

To call Lacaille's roster faint is no exaggeration. Only three — Circinus, Pictor, and Reticulum — have a star as bright as 3rd magnitude. Two — Mensa and Microscopium — have nothing brighter than 5th!

Toward the end of Lacaille's century, others decided to contribute more celestial clutter and came up with such gems as Machina Electrica; Officina Typographica, the Printing Shop; Globus Aerostaticus, the Hot-Air Balloon; and others that can be found on late 18th- and early 19th-century charts. All this was a celestial celebration of the era's high-tech wonders.

Mercifully, none of these post-Lacaille creations have survived. The fact that even Lacaille's junky constellations remained on the roster is perhaps due to his reputation as a scientist, individual, and even a nonpracticing theologian who had the title of abbot. He was a very skilled, energetic, and productive observer who created a remarkable catalog of nearly 10,000 stars and 42 deep-sky objects using a tiny 8-power refractor with a half-inch aperture.

Yet Lacaille's constellations have not been universally admired. The noted American astronomer Heber D. Curtis remarked that when he went to Chile before World War I to observe the southern heavens, he thought Lacaille had made those skies seem "like somebody's attic!"

— George Lovi

INTERESTING OBJECTS

Object	Constellation	Type	Right Ascension	Dec.	Size/Sep./ Period	Magnitude
M2	Aquarius	GC	$21^h\ 33^m.5$	–00° 49′	13′	6.5
M30	Capricornus	GC	$21^h\ 40^m.4$	–23° 11′	11′	7.5
NGC 7009 (Saturn Nebula)	Aquarius	PN	$21^h\ 04^m.2$	–11° 22′	2′	8
δ Apodis	Apus	DS	$16^h\ 20^m.3$	–78° 42′	103″	4.7, 5.1
o Ceti (Mira)	Cetus	VS	$02^h\ 19^m.3$	–02° 59′	332 days	2.0–10.1
30 Doradus (Tarantula Nebula, NGC 2070)	Dorado	DN	$05^h\ 38^m.7$	–69° 06′	40′ × 25′	5
θ Eridani (Acamar)	Eridanus	DS	$02^h\ 58^m.3$	–40° 18′	8.″2	3.4, 4.5
R Horologii	Horologium	VS	$02^h\ 53^m.9$	–49° 53′	404 days	4.7–14.3
β Tucanae	Tucana	DS	$00^h\ 31^m.5$	–62° 58′	27″.1	4.4, 4.8
47 Tucanae (NGC 104)	Tucana	GC	$00^h\ 24^m.1$	–72° 05′	30′.9	4
κ Pavonis	Pavo	VS	$18^h\ 56^m.9$	–67° 14′	9.1 days	3.94–4.75

DN = diffuse nebula; DS = double or multiple star; DkN = dark nebula; G = galaxy; GC = globular cluster; OC = open cluster; PN = planetary nebula; VS = variable star

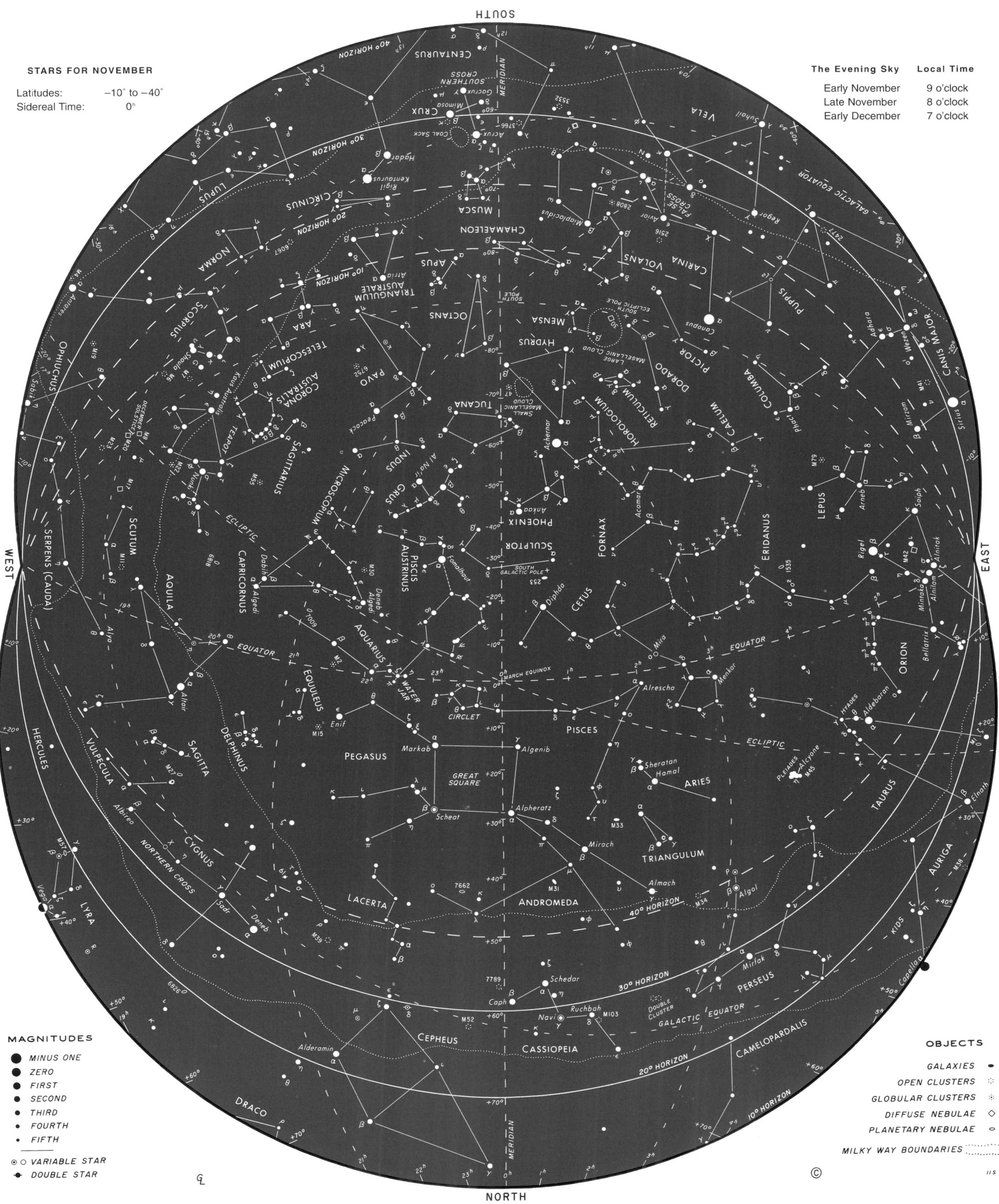

STARS FOR NOVEMBER
Latitudes: −10° to −40°
Sidereal Time: 0h
The Evening Sky
Local Time
Early November 9 o'clock
Late November 8 o'clock
Early December 7 o'clock
SOUTH
NORTH
EAST
WEST
MERIDIAN
MAGNITUDES
MINUS ONE
ZERO
FIRST
SECOND
THIRD
FOURTH
FIFTH
VARIABLE STAR
DOUBLE STAR
OBJECTS
GALAXIES
OPEN CLUSTERS
GLOBULAR CLUSTERS
DIFFUSE NEBULAE
PLANETARY NEBULAE
MILKY WAY BOUNDARIES
CENTAURUS
CRUX
SOUTHERN CROSS
VELA
LUPUS
CIRCINUS
MUSCA
FALSE CROSS
CHAMAELEON
CARINA
VOLANS
NORMA
APUS
TRIANGULUM AUSTRALE
PUPPIS
SCORPIUS
ARA
OCTANS
MENSA
CANIS MAJOR
OPHIUCHUS
TELESCOPIUM
PAVO
HYDRUS
DORADO
PICTOR
COLUMBA
CORONA AUSTRALIS
TUCANA
RETICULUM
HOROLOGIUM
CAELUM
SAGITTARIUS
INDUS
MICROSCOPIUM
GRUS
PHOENIX
LEPUS
SCUTUM
SERPENS (CAUDA)
SCULPTOR
FORNAX
ERIDANUS
AQUILA
CAPRICORNUS
PISCIS AUSTRINUS
CETUS
ORION
AQUARIUS
EQUULEUS
WATER JAR
CIRCLET
PISCES
ECLIPTIC
EQUATOR
MARCH EQUINOX
HERCULES
VULPECULA
SAGITTA
DELPHINUS
PEGASUS
GREAT SQUARE
ARIES
TAURUS
CYGNUS
NORTHERN CROSS
TRIANGULUM
LYRA
LACERTA
ANDROMEDA
AURIGA
PERSEUS
CEPHEUS
CASSIOPEIA
CAMELOPARDALIS
DRACO
GALACTIC EQUATOR
SOUTH POLE
SOUTH ECLIPTIC POLE
SOUTH GALACTIC POLE
LARGE MAGELLANIC CLOUD
SMALL MAGELLANIC CLOUD
COAL SACK
HYADES
PLEIADES
DOUBLE CLUSTER
KIDS
10° HORIZON
20° HORIZON
30° HORIZON
40° HORIZON
Hadar
Rigil Kentaurus
Acrux
Gacrux
Mimosa
Suhail
Regor
Avior
Miaplacidus
Canopus
Achernar
Ankaa
Diphda
Fomalhaut
Peacock
Al Na'ir
Antares
Shaula
Kaus Australis
Nunki
Acamar
Rigel
Sirius
Adhara
Wezen
Mirzam
Phact
Arneb
Saiph
Alnitak
Alnilam
Mintaka
Bellatrix
Mira
Menkar
Alrescha
Aldebaran
Alcyone
Elnath
Capella
Mirfak
Algol
Almach
Mirach
Hamal
Sheratan
Alpheratz
Algenib
Markab
Scheat
Enif
Altair
Alya
Dabih
Algedi
Deneb Algedi
Albireo
Sadr
Deneb
Vega
Schedar
Caph
Ruchbah
Navi
Alderamin

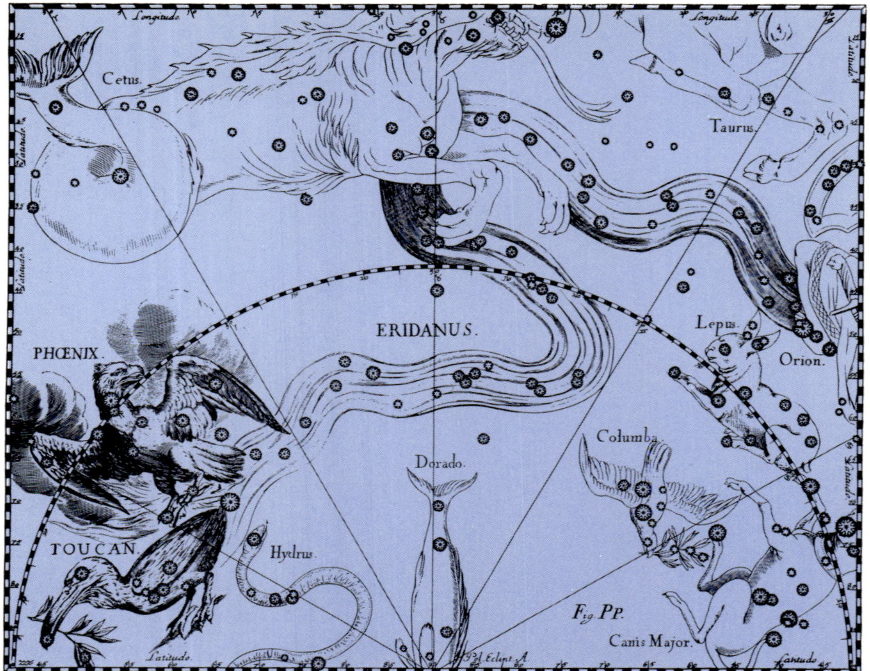

DECEMBER

BOATING DOWN THE ERIDANUS

WHILE skywatchers well north of the equator have the beauties of Perseus and Cassiopeia to admire high in December's early-evening sky, at midsouthern latitudes the meridian is almost bare of bright stars. The Milky Way curls around the horizon, so when we look toward the zenith we gaze almost directly out of the plane of our galaxy.

The sparsely populated River Eridanus occupies this region, extending over some 36^h of right ascension and almost 60° of declination. Although it is not often grouped with such easily recognized patterns as Scorpius, Crux, and the Big Dipper, I personally feel this meandering celestial river has to be one of the most aptly named constellations.

All of the ancient Western civilizations saw this constellation as a river — sometimes the Rhine or Rhone, sometimes the Po of modern Italy. In one myth it was the Eridanus, into which the youthful Phaethon was thrown when struck by Jupiter's thunderbolt. Alas, poor Phaethon's only crime was his inability to control the sun chariot entrusted to him for a day. His distraught sisters cried bitterly, and their tears fell into the Eridanus, where they were changed into amber (most appropriate, if we recall that the Po was the center of the amber trade in ancient times).

But Eridanus was perhaps most frequently associated with the Nile of Egypt. The ancient Egyptians knew it as *Al Nahr* (from which we get the word *Nile*) and found the Nile's unknown source well-symbolized by the disappearance of Eridanus below the southern horizon.

INTERESTING OBJECTS

Object	Constellation	Type	Right Ascension	Dec.	Size/Sep./ Period	Magnitude
Hyades	Taurus	OC	$04^h\ 27^m$	+16° 00′	6°	0.5
Large Magellanic Cloud	Dorado	G	$05^h\ 23^m.6$	–69° 45′	11° × 9°	0.6
Small Magellanic Cloud	Tucana	G	$00^h\ 52^m.7$	–72° 50′	5° × 3°	2.8
M33 (Pinwheel Galaxy)	Triangulum	G	$01^h\ 33^m.9$	+30° 39′	62′ × 39′	5.7
M45 (Pleiades)	Taurus	OC	$03^h\ 47^m.0$	+24° 07′	110′	1.2
M79	Lepus	GC	$05^h\ 46^m.7$	–24° 33′	9′	8.0
NGC 253 (Sculptor Galaxy)	Sculptor	G	$12^h\ 43^m.7$	+11° 33′	7′ × 6′	8.8
NGC 1535	Eridanus	PN	$04^h\ 14^m.2$	–12° 44′	0′.7	10
γ Arietis	Aries	DS	$01^h\ 53^m.5$	+19° 18′	7″.8	4.8, 4.8
o Ceti (Mira)	Cetus	VS	$02^h\ 19^m.3$	–02° 59′	332 days	2.0–10.1
β Doradus	Dorado	VS	$05^h\ 33^m.6$	–62° 29′	9.8 days	3.5–4.1
30 Doradus (Tarantula Nebula, NGC 2070)	Dorado	DN	$05^h\ 38^m.7$	–69° 06′	40′ × 25′	5
θ Eridani (Acamar)	Eridanus	DS	$02^h\ 58^m.3$	–40° 18′	8″.2	3.4, 4.5
β Tucanae	Tucana	DS	$00^h\ 31^m.5$	–62° 58′	27″.1	4.4, 4.8
47 Tucanae (NGC 104)	Tucana	GC	$00^h\ 24^m.1$	–72° 05′	30′.9	4
γ Volantis	Volans	DS	$07^h\ 08^m.8$	–70° 30′	13″.6	4.0, 5.9

DN = diffuse nebula; DS = double or multiple star; DkN = dark nebula; G = galaxy; GC = globular cluster; OC = open cluster; PN = planetary nebula; VS = variable star

Let's take a ride down Eridanus by first looking for its easy-to-find northern end, close to Rigel in Orion. From the 3rd-magnitude star Beta (β) Eridani the river flows first westward through Mu (μ) and Nu (ν) to Omicron^1 (o^1) and Omicron^2 (o^2). The latter (more popularly known as 40 Eridani from its numbering in Flamsteed's 1725 *Historia Coelestis Britannica*) is of special interest to us, for not only is it a fine telescopic triple star, but at a distance of 16 light-years it is also one of our closest neighbors — a distinction shared by several stars in this constellation. The brightest component is a 4th-magnitude orange-yellow dwarf. A 9½-magnitude secondary about 80 arc seconds away was discovered by Sir William Herschel in 1783, and an 11th-magnitude companion to this was first seen by Otto Struve in 1851.

The relationship of these two fainter stars is quite remarkable. The former, 40 Eridani B, is a white dwarf, while its companion, separated from it by 9 arc seconds, is a red-dwarf flare star. 40 Eridani B is the only white dwarf you can easily see in a small telescope (given that the well-known companion of Sirius is buried in the glare of its more prominent primary). With a diameter only about twice that of the Earth, yet a mass half that of the Sun, an average cubic inch of 40 Eridani B's material would weigh a whopping *two tons* here on Earth!

M79 in Lepus

The Hyades in Taurus

From 40 Eridani the constellation continues westward to Gamma (γ), Pi (π), and Delta (δ), then to magnitude-3.7 Epsilon (ε) Eridani. At a distance of 10.7 light-years this is another of our stellar neighbors; it is, in fact, the third-closest naked-eye star system after Alpha Centauri and Sirius. From Epsilon our river flows through Zeta (ζ) and Eta (η), and then briefly to Pi Ceti just inside the border of Cetus, the Sea Monster. (In ancient maps Cetus is usually depicted as resting on the bank of Eridanus with his forepaws in the water.) Turning southeast, we then meander back across 30° of sky before charting a course southwest to 3rd-magnitude Theta. This is a fine double star with components separated by 8.2 arc seconds and differing in brightness by about one magnitude. E. J. Hartung described them well when he said, "This brilliant white pair is one of the gems of the southern sky."

In ancient times Theta Eridani was taken as the southern end of the constellation, since points farther south never rose above the horizon from the latitudes of Arabia. However, modern astronomers have extended Eridanus another 18° south to magnitude-0.5 Achernar. Unhappily, Achernar's great southern declination of –57° means that its magnificence to Southern Hemisphere viewers is lost to those in the more populous North.

Because Eridanus contains few bright stars it is often overlooked in favor of its brighter surrounding attention-seekers. However, taking a celestial boat ride can be a relaxing and therapeutic experience. Try it sometime!

— *Graham Blow*

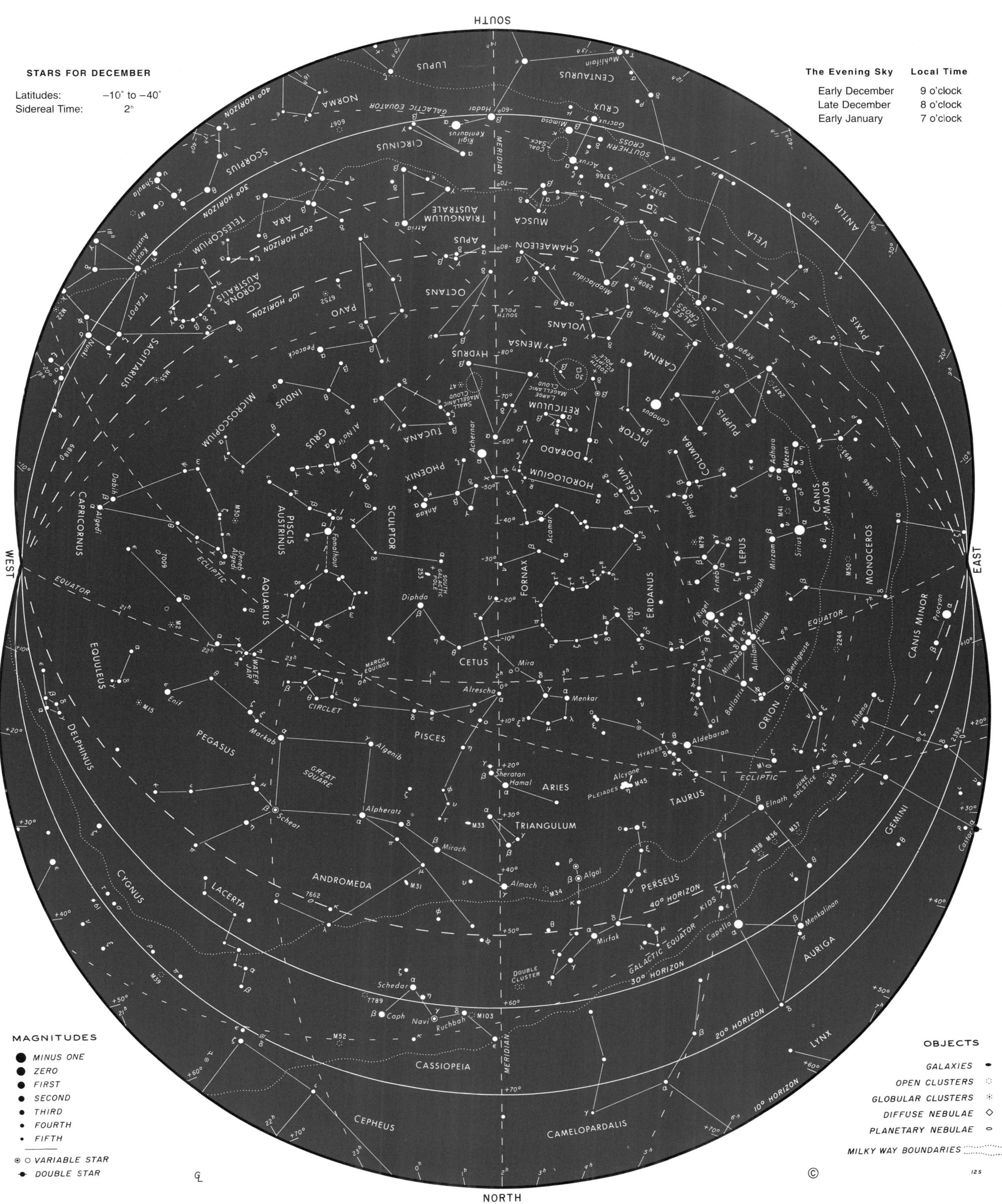
STARS FOR DECEMBER
Latitudes: −10° to −40°
Sidereal Time: 2h
The Evening Sky
Local Time
Early December 9 o'clock
Late December 8 o'clock
Early January 7 o'clock
SOUTH
NORTH
EAST
WEST
MAGNITUDES
MINUS ONE
ZERO
FIRST
SECOND
THIRD
FOURTH
FIFTH
VARIABLE STAR
DOUBLE STAR
OBJECTS
GALAXIES
OPEN CLUSTERS
GLOBULAR CLUSTERS
DIFFUSE NEBULAE
PLANETARY NEBULAE
MILKY WAY BOUNDARIES
CETUS
PISCES
PEGASUS
ANDROMEDA
ARIES
TRIANGULUM
TAURUS
ORION
PERSEUS
AURIGA
GEMINI
CASSIOPEIA
CEPHEUS
CAMELOPARDALIS
LYNX
ERIDANUS
FORNAX
SCULPTOR
AQUARIUS
PISCIS AUSTRINUS
CAPRICORNUS
LEPUS
CANIS MAJOR
CANIS MINOR
MONOCEROS
COLUMBA
PUPPIS
CARINA
PHOENIX
HOROLOGIUM
DORADO
RETICULUM
TUCANA
GRUS
INDUS
MICROSCOPIUM
SAGITTARIUS
PAVO
OCTANS
HYDRUS
MENSA
VOLANS
CHAMAELEON
APUS
MUSCA
TRIANGULUM AUSTRALE
ARA
TELESCOPIUM
CORONA AUSTRALIS
SCORPIUS
NORMA
LUPUS
CENTAURUS
CRUX
CIRCINUS
VELA
PYXIS
ANTLIA
PICTOR
CAELUM
EQUULEUS
DELPHINUS
CYGNUS
LACERTA
ECLIPTIC
EQUATOR
MERIDIAN
GALACTIC EQUATOR
SOUTH POLE
Sirius
Canopus
Achernar
Rigel
Betelgeuse
Aldebaran
Capella
Procyon
Fomalhaut
Diphda
Mira
Menkar
Alpheratz
Scheat
Markab
Algenib
Mirach
Almach
Algol
Mirfak
Hamal
Sheratan
Alcyone
PLEIADES
HYADES
Elnath
Menkalinan
Castor
Alhena
Bellatrix
Mintaka
Alnilam
Alnitak
Saiph
Arneb
Mirzam
Wezen
Adhara
Schedar
Caph
Ruchbah
Navi
Enif
Algedi
Dabih
Deneb Algedi
Ankaa
Al Na'ir
Peacock
Nunki
Kaus Australis
Shaula
Hadar
Rigil Kentaurus
Mimosa
Acrux
Gacrux
Atria
Miaplacidus
Avior
Suhail
Regor
Acamar
Alrescha
GREAT SQUARE
CIRCLET
WATER JAR
MARCH EQUINOX
JUNE SOLSTICE
KIDS
DOUBLE CLUSTER
SOUTHERN CROSS
FALSE CROSS
COAL SACK
LARGE MAGELLANIC CLOUD
SMALL MAGELLANIC CLOUD
SOUTH GALACTIC POLE
SOUTH GALACTIC POLE
10° HORIZON
20° HORIZON
30° HORIZON
40° HORIZON

Appendix A

The Greek Alphabet

α Alpha
β Beta
γ Gamma
δ Delta
ε Epsilon
ζ Zeta
η Eta
θ Theta
ι Iota
κ Kappa
λ Lambda
μ Mu
ν Nu
ξ Xi
ο Omicron
π Pi
ρ Rho
σ Sigma
τ Tau
υ Upsilon
φ Phi
χ Chi
ψ Psi
ω Omega

Note: These are the lowercase letters.

Appendix B

The 88 Constellation Names

Abbr.	Name	Latin Genetive	Meaning
And	Andromeda	Andromedae	Andromeda
Ant	Antlia	Antliae	Air Pump
Aps	Apus	Apodis	Bird of Paradise
Aqr	Aquarius	Aquarii	Water Bearer
Aql	Aquila	Aquilae	Eagle
Ara	Ara	Arae	Altar
Ari	Aries	Arietis	Ram
Aur	Auriga	Aurigae	Charioteer
Boö	Boötes	Boötis	Herdsman
Cae	Caelum	Caeli	Engraving Tool
Cam	Camelopardalis	Camelopardalis	Giraffe
Cnc	Cancer	Cancri	Crab
CVn	Canes Venatici	Canum Venaticorum	Hunting Dogs
CMa	Canis Major	Canis Majoris	Big Dog
CMi	Canis Minor	Canis Minoris	Little Dog
Cap	Capricornus	Capricorni	Sea Goat
Car	Carina	Carinae	Ship's Keel
Cas	Cassiopeia	Cassiopeiae	Queen Cassiopeia
Cen	Centaurus	Centauri	Centaur
Cep	Cepheus	Cephei	King Cepheus
Cet	Cetus	Ceti	Whale
Cha	Chameleon	Chamaeleontis	Chameleon
Cir	Circinus	Circini	Drawing Compass
Col	Columba	Columbae	Dove
Com	Coma Berenices	Comae Berenices	Berenice's Hair
CrA	Corona Australis	Coronae Australis	Southern Crown
CrB	Corona Borealis	Coronae Borealis	Northern Crown
Crv	Corvus	Corvi	Crow
Crt	Crater	Crateris	Cup
Cru	Crux	Crucis	Southern Cross
Cyg	Cygnus	Cygni	Swan
Del	Delphinus	Delphini	Dolphin
Dor	Dorado	Doradus	Goldfish
Dra	Draco	Draconis	Dragon
Equ	Equuleus	Equulei	Little Horse
Eri	Eridanus	Eridani	River Eridanus
For	Fornax	Fornacis	Furnace
Gem	Gemini	Geminorum	Twins
Gru	Grus	Gruis	Crane
Her	Hercules	Herculis	Hercules
Hor	Horologium	Horologii	Clock
Hya	Hydra	Hydrae	Sea Serpent
Hyi	Hydrus	Hydri	Water Snake
Ind	Indus	Indi	Indian
Lac	Lacerta	Lacertae	Lizard

Abbr.	Name	Latin Genetive	Meaning
Leo	Leo	Leonis	Lion
LMi	Leo Minor	Leonis Minoris	Little Lion
Lep	Lepus	Leporis	Hare
Lib	Libra	Librae	Scales
Lup	Lupus	Lupi	Wolf
Lyn	Lynx	Lyncis	Lynx
Lyr	Lyra	Lyrae	Lyre
Men	Mensa	Mensae	Table Mountain
Mic	Microscopium	Microscopii	Microscope
Mon	Monoceros	Monocerotis	Unicorn
Mus	Musca	Muscae	Fly
Nor	Norma	Normae	Level
Oct	Octans	Octantis	Octant
Oph	Ophiuchus	Ophiuchi	Snake Holder
Ori	Orion	Orionis	Orion
Pav	Pavo	Pavonis	Peacock
Peg	Pegasus	Pegasi	Pegasus
Per	Perseus	Persei	Perseus
Phe	Phoenix	Phoenicis	Phoenix
Pic	Pictor	Pictoris	Easel
Psc	Pisces	Piscium	Fishes
PsA	Piscis Austrinus	Piscis Austrini	Southern Fish
Pup	Puppis	Puppis	Ship's Stern
Pyx	Pyxis	Pyxidis	Ship's Compass
Ret	Reticulum	Reticuli	Eyepiece Reticle
Sge	Sagitta	Sagittae	Arrow
Sgr	Sagittarius	Sagittarii	Archer
Sco	Scorpius	Scorpii	Scorpion
Scl	Sculptor	Sculptoris	Sculptor's Apparatus
Sct	Scutum	Scuti	Shield
Ser	Serpens	Serpentis	Serpent
Sex	Sextans	Sextantis	Sextant
Tau	Taurus	Tauri	Bull
Tel	Telescopium	Telescopii	Telescope
Tri	Triangulum	Trianguli	Triangle
TrA	Triangulum Australe	Trianguli Australis	Southern Triangle
Tuc	Tucana	Tucanae	Toucan
UMa	Ursa Major	Ursae Majoris	Big Bear
UMi	Ursa Minor	Ursae Minoris	Little Bear
Vel	Vela	Velorum	Ship's Sails
Vir	Virgo	Virginis	Virgin
Vol	Volans	Volantis	Flying Fish
Vul	Vulpecula	Vulpeculae	Little Fox

INDEX

Boldface numbers indicate photographs. *Italic* numbers indicate entries in tables of interesting objects.

Aberration, of starlight, 28
Achernar, 62
Aesculapius, 22
Al Jauzah, 42
Alcor, 20
Aldebaran, 12
Algieba, 20
Algol, 30
Alkaid, 18
Almagest, 56
Alpha Canum Venaticorum, *18, 20*
Alpha Capricorni, *28, 56,* 58
Alpha Centauri, 7, 28, 46, **46,** 48, *48,* 50, *50,* 62
Alpha Crucis, *46,* 48, *48, 50*
Alpha Geminorum, *14, 16*
Alpha Gruis, 58
Alpha Herculis, *24*
Alpha Librae, *50*
Alpha Orionis, *14, 42*
Alphecca, 18
Alula Australis, 18
Alula Borealis, 18
American Association of Variable Star Observers (AAVSO), 30
Andromeda, 16, 20, 32, 40
Andromeda Galaxy, 20
Antares, 56
Antlia, 60
Apus, 60
Aquarius, 58
Ara, 26
Aratus, 16
Arc minutes, 6, 8
Arc seconds, 6, 8
Argo Navis, 42, 44
Aries, 16
Associations, 18, 20
Auriga, 16

Baade, Walter, 40
Bayer, Johann, 7, 52, 58, 60
Beehive Cluster, 48
Beginner's Guide to the Skies, The, 26
Bellerophon, 32
Bessel, Friedrich Wilhelm, 28
Beta Carinae, 48
Beta Centauri, 46, **46,** 48, 50, 52
Beta Crucis, 46
Beta Cygni, *28*
Beta Doradus, *40, 62*
Beta Eridani, 62
Beta Gruis, 58
Beta Lyrae, *28*
Beta Persei, *12, 30, 32, 40*
Beta Scorpii, *24,* 56
Beta Tucanae, *58, 60, 62*
Betelgeuse, 14, 18
Big Dipper, 6, 18, 20, 56, 62
Blow, Graham, 4
Bok, Bart, 40
Bonaparte, Napoleon, 32
Bradley, James, 28
Braymer, Lawrence, 34
BS 4768, 46

Caelum, 60
Cancer, 48
Canis Major, 18
Canopus, 42
Capella, 16
Capricornus, 32, *58*
Carina, 40, 42, 44, 46
Cassiopeia, 16, 20, 32, 62
Centaurus, 7, 20, 32, 40, 46, 50
Cepheus, 16, 32
Cetus, 16, 30, 58, 62
Chamaeleon, 60
Chimera, the, 32
Chiron, 50
Circinus, 20, 60
Cleminshaw, C. H., 26
Clusters, galaxy, 20; globular, 20, 52; open, or galactic, 12, 20, 44
Coalsack Nebula, 40, *46, 48,* **48,** *50, 52*
Color temperature, 14
Coma Berenices, 12, 20
Coma Cluster, *18, 20, 22, 48, 50*
Compass points, 6
Constellations, names of, 65
Coordinates, celestial, 6, 7
Corona Borealis, 18, 20
Crab Nebula, 12,
Crux, 46, 48, 62
Curtis, Heber D., 60
61 Cygni, 28, *28*

Dante, 46
Dark adaptation, 6
Declination (Dec.), 6
Deep-sky objects, 7; symbols on charts, 7
Degrees, 6
Delphinus, 46
Delta Apodis, *56, 58, 60*
Delta Centauri, 50
Delta Cephei, 30, *30, 32,* 48
Delta Eridani, 62
Delta Velorum, 48
Deneb, 28
Diamond Cross, 48
Dnoces, 18
Dorado, 42, 60
30 Doradus, *42, 44, 46, 58, 60, 62*
Double stars, optical, 7; separation of, 8; symbol on charts, 7
Draco, 16, 32
Dreyer, Johann, 7
Dubhe, 18

Eclipsing binaries, 30
Ecliptic, 6
Elements, formation, 14
Enif, 32
Epsilon Boötis, 20, *20, 22, 24*
Epsilon Carinae, 44, 48
Epsilon Centauri, 50, 52
Epsilon Crucis, 46
Epsilon Cygni, 28
Epsilon Eridani, 62
Equator, celestial, 5
40 Eridani, 62
40 Eridani B, 62
Eridanus, 58, 62
Eta Carinae, 40, 44, *44,* **44,** *46,* 48, *48*
Eta Eridani, 62
Eta Herculis, 52
Eta Scorpii, 56
Eudoxus, 16

False Cross, 48
Field Guide to the Stars and Planets, 46
Flamsteed, John, 12, 62
Flare star, 50, 58
Flashlight, red, 6
Fomalhaut, 58
Fornax, 60

Gaia, 56
Galactic equator, 7
Galactic plane, 26
Galaxies, 20
Gamma Andromedae, *12, 32*
Gamma Arietis, *32, 34, 62*
Gamma Cassiopeiae, *30, 32*
Gamma Centauri, 50
Gamma Crucis, 48
Gamma Eridani, 62
Gamma Gruis, 58
Gamma Leonis, *16, 18,* 20, *20*
Gamma Velorum, 44
Gamma Virginis, 20, *50*
Gamma Volantis, *62*
Georgian Star, 12
Gilgamesh, 24
Gould, Benjamin Apthorp, 40
Gould's Belt, 40
Gravitation, 20
Great Star of the South, 42
Great Bear, 18
Greek letters, 7, 64
Grus, 58, 60

Halley, Edmund, 28, 52
Hartung, E. J., 62
Heliometer, 28
Henderson, Thomas, 28
Heracles, 24, 26
Hercules , 16, 20, 24, 32, 52
Herschel, John, 46, 50
Herschel, William, 12, 28, 62
Hippocrates, 22
Hippolytus, 22
Historia Coelestis Britannica, 62

Hood, Thomas, 18
Horizon, on charts, 5
Hour circles, 6
Hyades, 12, *12, 34, 40, 42, 62,* **62**
Hydra, 32
Hydrogen, fusing, 14
Hydrus, 60
Hygeia, 22

IC 2391, 44
IC 2602, 44
Indus, 60
Iota Carinae, 48
Iota Ursae Majoris, 18
Ishtar, 24
Izar, 20

Jewel Box Cluster, 46, *46, 48, 50*

Kappa Crucis, 46, *46, 48, 50*
Kappa Pavonis, *58, 60*
Kappa Ursae Majoris, 18
Kappa Velorum, 48
Keyser, Pieter Dirckszoon, 60

l Carinae, *44, 46,* 48, *48*
Lacaille, Nicolas Louis de, 60, **60**
Lagoon Nebula, 54
Lambda Centauri, 50
Lambda Sagittarii, 52
Lambda Scorpii, 56
Lambda Ursae Majoris, 18
Lambda Velorum, 42, 44
Large Magellanic Cloud, 40, *40,* **40,** *42, 44, 58, 62*
Latitude, determining, 7
Leavitt, Henrietta, 40
Leo, 16, 20
Light pollution, 6
Little Dipper, 12
Local Group, 20
Lovi, George, 4, 7

M1, *12,* **12,** *14, 40, 42*
M2, *28, 30, 56, 58, 60*
M3, *20,* **20,** *22, 50*
M4, *24, 52,* 56
M5, *22,* **22,** 24, *50, 52*
M6, *26,* 48, *52, 54,* 56, *56,* **56**
M7, *26, 52, 54,* 56, *56,* **56**
M8, *26, 52,* 54, *54,* **54,** *56*
M10, *24,* 26, *52, 54*
M11, *26, 28, 54, 56*
M12, *24, 26,* 52, *54*
M13, 20, *24,* **24,** *26,* 52
M15, *28, 30, 56, 58*
M17, *26,* **26,** *54, 56*
M19, *24,* 26, *52, 54*
M20, *26, 52,* 54, *54,* **54,** *56*
M22, *26,* 52, *52,* **52,** *54*
M23, *26,* 54, *54*
M27, *28,* **28**
M30, *30, 56, 58,* **58,** *60*
M31, 20, *30, 32,* 40
M33, 20, *32,* **32,** *34, 62*
M34, *12, 34,* **34,** *40*
M35, *14, 16,* **16**
M36, *12, 14, 16*
M37, *12, 14,* **14,** *16*
M38, *12, 14, 16*
M39, *28,* **28,** *30, 32*
M41, *14, 42,* **42,** *44*
M42, *12, 14,* **14,** *40, 42*
M44, *16,* **16,** *18, 44, 46*
M45, 7, *12,* **12,** *34, 40, 62*
M46, *16, 42, 44,* **44**
M50, *14, 16, 42, 44*
M51, *20,* **20,** *22*
M52, *30, 32, 34*
M53, *20, 22, 48, 50,* **50**
M55, *54,* **54,** *56, 58*
M57, *26, 28, 54*
M64, *20, 22, 48,* **48,** *50*
M67, *16, 18, 44, 46*
M79, *14, 40, 42, 62,* **62**
M80, *24, 52*
M81, *12, 14, 16, 18,* **18,** *20, 22*
M82, *12, 14, 16, 18, 20, 22*
M92, *24, 26*
M93, *16, 42, 44,* **42**
M94, *18, 20, 22*
M103, *12, 30,* **30,** *32, 34*
M104, *20, 22,* **22,** *48, 50*
M106, *18, 20, 22,* **22**
Maan, 42
Magellanic Clouds, 20, 40, 58
Magnitude, visual, 7
Main-sequence stars, 14
Maori, 42
Markab, 6
Maui, Hawaii, 56
Mensa, 60
Menzel, Donald, 46
Meridian, 6, 22
Messier, Charles, 7, 54
Messier objects, 7
Microscopium, 60
Milky Way, 7, 12, 20, 26, 34, 40, 42, 44, 48, 52, 54, 58, 60, 62; band, 20, 26; Orion arm, 12; spiral arms, 12
Minerva, 32
Mira, 30, 58; estimating brightness of, 30
Mizar, 20
Monoceros, 32
Moon illusion, 6, 32
Mu Eridani, 62
Mu Scorpii, 56, *56*
Mu Ursae Majoris, 18
Multiple-star systems, 20; designation, 8
Musca, 60
Muscida, 18

Nebulae, 44
New General Catalogue of Nebulae and Clusters of Stars, 7
New Patterns in the Sky, The, 22
NGC objects, 7
NGC 246, 58
NGC 253, *32, 34,* 58, *58,* **58,** *62*
NGC 869/884, *12, 32,* **32,** *34*
NGC 1535, *12, 34, 40, 62*
NGC 2244, *14, 42*
NGC 2392, *14, 16, 44,* **44**
NGC 2477, *42, 44*
NGC 2516, *42,* 44, *44, 46,* 48
NGC 2808, *44, 48*
NGC 3132, *46, 48*
NGC 3242, *18, 20, 46,* **46,** *48*
NGC 3372, 44, *44,* **44,** 46, 48
NGC 3532, *44, 46,* 48
NGC 3766, *46, 48,* 50, *50*
NGC 4755, 46, *46, 48, 50*
NGC 5128, *50*
NGC 5139, *48, 50, 52,* **52**
NGC 6067, *50, 52*
NGC 6210, *24, 26, 52, 54*
NGC 6231, 56
NGC 6543, 18, *22, 24,* **24,** *26, 28, 30*
NGC 6572, *54*
NGC 6752, *54, 56, 58*
NGC 6818, *28*
NGC 6826, *26, 28*
NGC 7009, *28, 30, 56, 58, 60,* **60**
NGC 7662, *32*
NGC 7789, *30, 32, 34,* **34**
Northern Cross, 28
Northern Hemisphere, 9–35
Nu Draconis, 24
Nu Eridani, 62
Nu Scorpii, 56
Nu Ursae Majoris, 18

Octans, 60
Omega Carinae, 48
Omega Centauri, 20, 40, *48,* 50, *50,* 52, *52,* **52**
Omicron Ceti, *32, 34, 40,* 58, *60, 62*
Omicron Velorum, 44
Omicron^{1} Eridani, 62
Omicron^{2} Eridani, 62
Ophiuchus, 22
Orion, 6, 14, 16, 18, 20, 22, 26, 40, 42, 44, 46, 56

Pan, 58
Panacea, 22
Papa, 42
Parallax, 28
Pasachoff, Jay M., 46
Pavo, 60
Pegasus, 16, 32; Great Square of, 6, 32
Perseus, 16, 30, 62
Phaenomena, 16
Phaethon, 62
Phoenix, 32, 60
Pi Centauri, 50
Pi Ceti, 62
Pi Eridani, 62
Pictor, 42, 60
Pisces, 6, 20, 58
Piscis Austrinus, 58
Planets, visible to naked eye, 6
Pleiades, 7, 12, 20, 44
Pluto, god, 22
Polaris (North Star), 7, 30, 60
Population I regions, 40
Population II regions, 40
Porrima, 20
Praesepe, 48
Precession, 6–7, 42, 60
Proctor, Richard A., 18
Procyon, 42
Proper motion, 28
Proxima Centauri, 50

Ptolemy, 42, 52, 56
Pulcherrima, 20
Puppis, 42, 44
Pyxis, 42, 44

R Coronae Borealis, *22*
R Horologii, *60*
R Lyrae, *28*
Rambling Through the Skies, 4
Rangi, 42
Realm of the Galaxies, 20
Red LED flashlight, 6
Regor, 44
Reticulum, 60
Rho Persei, *34*
Riccioli, Giambattista, 20
Rigel, 7, 14, 42
Right ascension (R.A.), 6
Royer, Augustin, 46

Sadr, 28
9 Sagittarii, 54
Sagittarius, 12, 40, 50, 52, 54
Sagittarius A*, 54
SAO 227392, 56
SAO 252073, 46
Scheat, 32
Schiller, Julius, 46
Scorpius, 12, 22, 26, 40, 44, 56, 62
Scorpius-Centaurus Association, 20, 56
Sculptor, 20, 58, 60
Sculptor Galaxy, 58
Seeing, 34
Seneca, Lucius Annaeus, 24
Serpens, 22
Shapley, Harlow, 40
Sigma Octantis, 60
Sirius, 6, 7, 16, 42, 44, 62
Sky & Telescope, 4, 5, 6, 30
Small Magellanic Cloud, 40, *40,* **40,** *42, 44,* 52, *56, 58,* 60, *62*
Smoky Way, 26
South celestial pole, 52, 60
Southern Cross, 6, 46, **46,** 48, 50, 56
Southern Hemisphere, 37–63
"Southern Pleiades," 44
Staal, Julius, 22
Star charts, how to use, 5–6; usage table, 5
Star colors, 14
Star distances, measuring, 28
Star magnitudes, 7
Star names and designations, 7
Star system, 50
Stars, blue supergiant, 14, 26; double, 7, 8, 50; naked-eye, 26; red dwarf, 14; red giant, 14; red supergiant, 14; variable, 7, 8, 30
Stereographic projection, 6
Struve, Friedrich Wilhelm, 20, 28
Struve, Otto, 62
Suhail, 42, 44
Sun, 20
Supernova, 14

Tane, 42
Tania Australis, 18
Tania Borealis, 18
34 Tauri, 12
Taurus, 7, 12, 16
Teapot asterism, 12, 26, 54
Telescopium, 60
Theta Carinae, 44, 48
Theta Centauri, 50
Theta Eridani, *40, 60,* 62, *62*
Theta Scorpii, 56
Tirion, Wil, 4
Transparency, atmospheric, 34
Triangulum Australe, 60
Triangulum spiral galaxy, 20
Trifid Nebula, 54
Tube currents, 32
Tucana, 20, 60
47 Tucanae, *40,* 52, *52,* **52,** *56, 58, 60, 62*
Turbulence, atmospheric, 34
Typhon, 58

Upsilon Carinae, 48
Upsilon Scorpii, 56
Uranometria, 52, 60
Uranus, 12
Ursa Major, 16, 18
Ursa Major stream, 18
Utnapishtim, 24
UV Ceti, 58

ν **Draconis,** *24*
Variable stars, 7, 8, 30; Cepheid, 30, 40, 48; period, 8; symbol on charts, 30
Vega, 28
Vela, 40, 42, 44
Vernal equinox, 6
Vespucci, Amerigo, 46
Virgo, 20, 22
Virgo Cloud, 20

Werner, Helmut, 16
White, Edward H., II, 18
Wolf-Rayet stars, 44

Xi Ursae Majoris, 18

Zenith, 6
Zeta Centauri, 50
Zeta Eridani, 62
Zeta Gruis, 58
Zeta Puppis, 44
Zeta Ursae Majoris, *18, 22, 24*
Zeta^1 Scorpii, 56
Zeta^2 Scorpii, 56
Zeus, 22, 26
Zodiac, 12